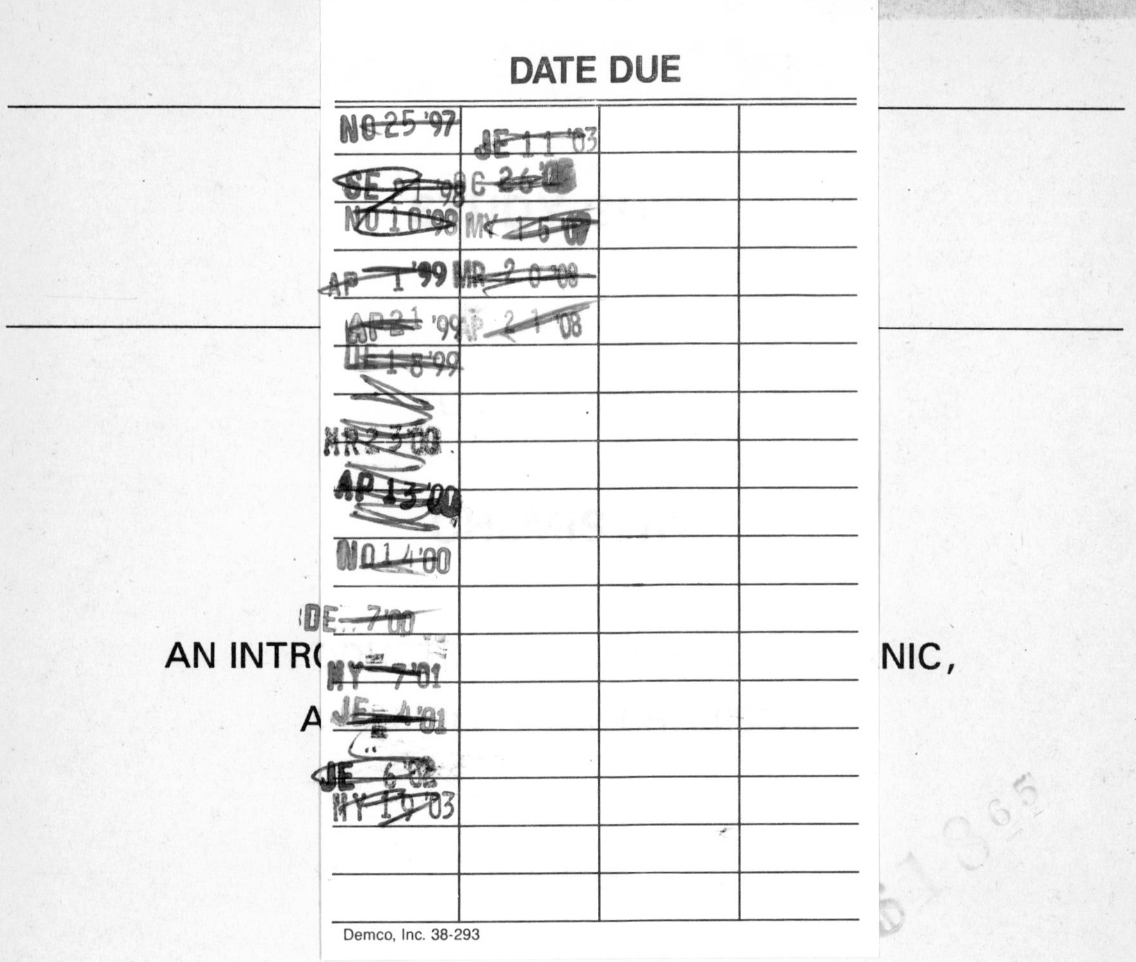

AN INTRO... NIC,

A...

KAREN C. TIMBERLAKE

Los Angeles Valley College

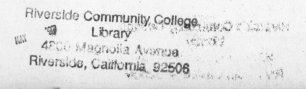

HarperCollins*CollegePublishers*

Study Guide to accompany CHEMISTRY: AN INTRODUCTION TO GENERAL, ORGANIC, AND BIOLOGICAL CHEMISTRY, Fifth Edition

ISBN: 0-06-046577-8

92 93 94 95 96 9 8 7 6 5 4 3

TABLE OF CONTENTS

INFORMATION FOR THE STUDENT

Now that you are ready to sit down and study your chemistry, let's go over some learning methods that can help you study chemistry. The materials in these chapters are designed to help you understand and practice the chemical concepts that are presented in class and in your text.

Study One Section at a Time and Practice Some Problems

Each section in this study guide contains a set of Learning Exercises to match each section in the text using a variety of questions. Answers are given for all of the Learning Exercises.

I often recommend a study system to students in which you read one section of the text and ***immediately practice some study problems*** that go with it. In this way, you concentrate on a small amount of information and actively use what you learned to answer questions. This helps you to organize and review the information without being overwhelmed by the entire chapter. It is important to understand each section, because they build like steps. Information presented in each chapter proceeds from the basic to the more complex skills. Some sections are designed to highlight certain areas which are particularly difficult for students. Perhaps you will only study 3 or 4 sections of the chapter. As long as you also practice some problems, the information will stay with you.

Use Writing To Learn Content

At various times, you will notice some essay questions that illustrate one of the concepts. I believe that writing out your ideas is a very important way of learning content. If you can put your problem-solving techniques into words, then you understand the patterns of your thinking and you will find that you have to memorize less.

Try the Self-Evaluation Tests and Evaluate Your Level of Mastery

A Self-evaluation Test is available at the end of each chapter. If you think that you have learned the material by studying your text, you might wish to use the practice test as a pretest. If the results of this pretest indicate that you know the material, you are ready to proceed to the next chapter. If, however, the results indicate further study is needed, you can repeat the practice problems. Answers for all of the problems as well as the self-evaluation exams are included at the end of each section. Each answer set may be covered up as you work the problems and then uncovered to check your results.

The completion of the Learning Exercises in each chapter will prepare you for your Self-Evaluation Test found at the end of each chapter. If you think that you have learned the material by studying your text and study guide, you may wish to use the self-evaluation test as a pretest. If the results of this pretest indicate that you know the material, you will wish to proceed to the next chapter. If, however, the results indicate further study is needed, you may want to return to the Learning Exercises and review the sections and problems again. All the answers are given to each Self-Evaluation Test immediately after each test. The successful results on the practice test will indicate your level of mastery. You can score the Self-Evaluation Test and determine your learning level.

Score	***Level of Mastery***
90-100	***Excellent.*** You show a high level of mastery of this material.
80-90	***Good.*** You have a good basic understanding of the material. Check your calculations and review your thinking in the problems that you missed. You may just need to be more careful or to review and practice one or two skill areas.
70-80	***Passing***. This is a passing level. You may want to review your text again and practice the kinds of problems which gave you some difficulty. This is a good time to see your instructor during office hours.
below 70	***Not Passing.*** This is not a mastery level of performance. Check your calculations again and go through the Learning Exercises once more. Then retake the Self-Evaluation Test.

Go to Your Teacher's Office Hours

Your teacher wants you to understand and enjoy learning this material and should have office hours. Don't be intimated. Going to see your teacher is one of the best ways to clarify what you need to learn in chemistry.

CHAPTER 1
MEASUREMENTS

KEY CONCEPTS

1. The meter is the metric unit for length; the gram for mass; the liter for volume; and Celsius degree or Kelvin for temperature.

2. The metric system is a decimal system; prefixes indicating multiples or fractions of ten are attached to the units of measurements to express different sizes of measurement.

3. An equality expresses a relationship between two units of measurement.

4. Significant figures are the numbers used to report a measurement. In calculations with measurements, the numerical answers must be adjusted for the correct number of significant figures.

5. A conversion factor is derived from an equality by writing the quantities in the relationship as a ratio.

6. In the problem-solving process, conversion factors are used to change a given unit to a desired unit.

7. The density of a substance is its mass per unit volume (D=m/v). The specific gravity of a substance is the ratio of its density to the density of water (1.00 g/mL).

KEY WORDS

Using *complete* sentences, write a description for each of the following key words:

metric system

significant figures

conversion factor

density

LEARNING EXERCISES

UNITS OF MEASUREMENT (1.1)

A. Match the metric terms with A. metric system B. meter C. liter D. gram

1._____length 2._____volume

3._____mass 4._____a decimal system of measurement

Answers 1. B 2. C 3. D 4. A

METRIC PREFIXES (1.2)

B. Match the items in column A with those from column B that match most closely.

A	B
1._____kilo-	a. ten grams
2._____one thousand liters	b. millimeter
3._____deciliter	c. one-tenth of a liter
4._____milliliter	d. one hundred liters
5._____centimeter	e. kiloliter
6._____one-tenth centimeter	f. one-hundredth of a meter
7._____hectoliter	g. 1000 g
8._____kilogram	h. one-thousandth of a liter
9._____dekagram	i. one thousand times

Answers 1. i 2. e 3. c 4. h 5. f 6. b 7. d 8. g 9. a

C. Number the items in the following sets from smallest (1) to largest (5):

1. kilogram ____ centigram____ hectogram____ milligram____ gram____
2. centimeter____ kilometer____ decimeter____ millimeter____ meter____
3. cL____ dL____ mL____ kL____ L____
4. kg____ mg____ dag____ dg____ cg____

Answers (1) 5, 2, 4, 1, 3 (2) 2, 5, 3, 1, 4 (3) 2, 3, 1, 5, 4 (4) 5, 1, 4, 3, 2

WRITING NUMERICAL RELATIONSHIPS (1.3)

D. Complete the following metric relationships:

1. 1 L	= _____mL		6. 1 L	= _____dL	
2. 1 m	= _____cm		7. 1 dL	= _____mL	
3. 1 kg	= _____g		8. 1 cm	= _____mm	
4. 1 mg	= _____μg		9. 1 dL	= _____L	
5. 1 m	= _____mm		10. 1 hm	= _____m	

Answers: (1) 1000 (2) 100 (3) 1000 (4) 1000 (5) 1000 (6) 10 (7) 100 (8) 10 (9) 0.1 (10) 100

SIGNIFICANT FIGURES (1.4)

E. State the number of significant figures in the following measured numbers:

1.	35.24 g	_____	5.	0.000080	_____
2.	55,000 m	_____	6.	805 mL	_____
3.	5.025 L	_____	7.	0.006 kg	_____
4.	268,200 mm	_____	8.	25.0°C	_____

Answers: (1) 4 (2) 2 (3) 4 (4) 4 (5) 2 (6) 3 (7) 1 (8) 3

F. Use a calculator to solve the following and give an answer with the correct number of significant figures:

1. $1.3 \times 71.5 =$

2. $\dfrac{8.00}{4.00} =$

3. $\dfrac{0.082 \times 25.4}{118.5} =$

4. $\dfrac{3.05 \times 1.86}{0.116 \times 3.4} =$

5. $4.2 + 8.15 =$

6. $38.520 - 11.4 =$

Answers: (1) 93 (2) 2.00 (3) 0.018 (4) 14 (5) 12.4 (6) 27.1

CONVERSION FACTORS (1.5)

Review: A conversion factor describes an equality as a ratio. For example, the equality 2.54 cm = 1 inch is written as the factors

$$\frac{2.54\ \text{cm}}{1\ \text{inch}} \quad \text{and} \quad \frac{1\ \text{inch}}{2.54\ \text{cm}}$$

G. Write two conversion factors for the following pairs of units:

1. millimeters and meters
2. milligrams and grams
3. kilograms and pounds
4. inches and centimeters
5. centimeters and meters
6. milliliters and quarts
7. deciliters and liters
8. millimeters and centimeters
8. millimeters and centimeters
10. kilograms and grams

Answers: (1) 1000 mm/m; 1 m/1000 mm (2) 1000 mg/g; 1 g/1000 mg
(3) 2.2 lb/kg; 1kg/2.2 lb (4) 2.54 cm/in; 1 in/2.54 cm (5) 100 cm/m; 1 m/100 cm
(6) 946 mL/qt; 1 qt/ 946 mL (7) 10 dL/L; 1L/10 dL (8) 10 mm/cm; 1 cm/ 10 mm
(9) 1.06 qt/L; 1 L/1.06 qt (10) 1000 g/kg; 1 kg/1000 g

PROBLEM SOLVING USING CONVERSION FACTORS (1.6)

H. Use conversion factors to solve the following METRIC-METRIC problems:

1. 189 mL = ______L
2. 2.7 cm = ______mm
3. 0.0025 L = ______mL
4. 76 mg = ______g

5. 576 mm = _____m   6. 28.5 mL = _____L

7. 1500 μg = _____g   8. 1.5 km = _____m

9. 82 m = _____cm   10. 0.241 L = _____mL

Answers: (1) 0.189 L (2) 27 mm (3) 2.5 mL (4) 0.076 g (5) 0.576 m
(6) 0.0285 L (7) 0.0015 g (8) 1500 m (9) 8200 cm (1) 241 mL

I. Use conversion factors to solve the following METRIC-AMERICAN problems:

1. 18 inches = _____cm   6. 150 lb = _____kg

2. 4.0 qts = _____L   7. 840 g = _____lb

3. 275 mL = _____qt   8. 15 ft = _____cm

4. 1300 mg = _____lb   9. 8.50 oz = _____g

5. 0.450 yd = _____mm   10. 88 mL = _____cups

Answers: (1) 46 cm (2) 3.8 L (3) 0.291 qt (4) 0.0029 lb (5) 411 mm
(6) 68 kg (7) 1.9 lb (8) 460 cm (9) 241 g (10) 0.37 cups

J. Use conversion factors to solve the following problems:

1. A piece of plastic tubing measures 120 mm. What is the length of the tubing in inches?

2. A statue weighs 240 pounds. What is the mass of the statue in kilograms?

3. A new patient in admitting has a height of 6 feet three inches. What is the patient's height in meters?

4. Sandee, a sculptor, has prepared a mold for casting a silver candlestick. The mold has a volume of 0.35 pints. If the price of silver is \$5.93 per mL, how much will the required silver cost in dollars?

5. A doctor has ordered 0.450 g of a sulfa drug. On hand are 150-mg tablets. How many tablets are needed?

Answers: (1) 4.7 inches (2) 110 kg (3) 1.9 m (4) \$980 (5) 3 tablets

DENSITY AND SPECIFIC GRAVITY (1.7)

Review: The density of a substance is the ratio of its mass and volume. For example, a liquid may have a density of 1.2 g/mL. This equality (1 mL = 1.2 g) can be used in a problem as a factor. If asked for the mass of 25 mL of the liquid, we can write

$$25 \text{ mL liquid} \quad \times \quad \frac{1.2 \text{ g}}{1 \text{ mL liquid}} \quad = \quad 30 \text{ mL}$$

If a metal with a mass of 48 g has a density of 8.4 g/mL, the volume is calculated as

$$48\text{ g} \times \frac{1\text{ mL}}{8.4\text{ g}} = 5.7\text{ mL}$$

K. Calculate density, or use density as a conversion factor to solve the following:

1. What is the density of 200.0 mL of glycerol if the sample has a mass of 252 g?

2. Diabetes insipidus is a disease whereby 5 to 12 liters of urine may be voided per day. Calculate the specific gravity of a 100.0 mL sample that has a mass of 100.2 g.

3. A small solid has a mass of 5.5 oz. When placed in a graduated cylinder with a water level of 25.2 mL, the object causes the water level to rise to 43.8 mL. What is the density of the object?

4. A sugar solution has a density of 1.20 g/mL. What is the mass in grams of 0.250 L of the solution?

5. Jerry, a jeweler, has a sample of gold that weighs 0.26 pounds. If gold has a density of 19.3 g/mL, what is the volume in mL of the gold sample?

6. A salt solution has a specific gravity of 1.15 and a volume of 425 mL. What is the mass in grams of the solution?

7. A 50.0-g sample of a glucose solution has a specific gravity of 1.28. What is the volume in liters of the sample?

Answers: (1) 1.26 g/mL (2) 1.002 (3) 8.4 g/mL (4) 300 g (5) 6.1 mL (6) 489 g (7) 0.0391 L

SELF-EVALUATION TEST

MEASUREMENTS

INSTRUCTIONS: Select the letter preceding the work or phrase that best completes or answers the question.

1. Which of the following is a metric measurement of volume?
A. kilogram B. kilowatt C. kiloliter D. kilometer E. kiloquart

2. Which of these prefixes is the largest?
A. deka- B. deci- C. hecto- D. kilo- E. micro-

3. What is the decimal equivalent of the prefix ***centi-***?
A. one-thousandth B. one-hundredth C. one-tenth D. ten E. one hundred

4. Which of the following is a conversion factor?
A. 12 inches B. 3 feet C. 20 meters
D. 12 inches/foot E. 2 cubic centimeters

5. Which is a conversion factor that relates milliliters to liters?
A. 1000 mL/L B. 100 mL/L C. 10 mL/L D. 0.01 mL/L E. 0.001 mL/L

6. Which is a conversion factor for millimeters and centimeters?
A. 1 mm/cm B. 10 mm/cm C.100 cm/mm
D. 100 mm/cm E. 10 cm/mm

7. Which of the following is the smallest unit of measurement?
A. gram B. milligram C. kilogram D. hectogram E. centigram

8. Which volume unit is the largest?
A. kg B. mm C. cm^3 D. L E. kL

9. 294 mm is equal to
A. 2940 m B. 29.4 m C. 2.94 m D. 0.294 m E. 0.0294 m

10. A tennis racket size is 4.5 inches. What is that size in centimeters?
A. 11 cm B. 1.8 cm C. 0.56 cm D. 450 cm E. 15 cm

11. What is the volume of 65 mL in liters?
A. 650 L B. 65 L C. 6.5 L D. 0.65 L E. 0.065 L

12. What is the mass of a 22 lb turkey?
A. 10 kg B. 48 kg C. 10,000 kg D. 0.048 kg E. 22,000 kg

13. The number of milliliters in 2 dekaliters is
A. 20 mL B. 200 mL C. 2000 mL
D. 20,000 mL E. 500,000 mL

14. A person who is 5 feet 4 inches tall would be
A. 64 m B. 25 m C. 14 m D. 1.6 m E. 1.3 m

15. How many ounces are in 1,500 grams? (1 lb = 16 oz)
A. 94 oz B. 53 oz C. 24,000 oz D. 33 oz E. 3.3 oz

16. How many quarts of orange juice are in 255 mL of juice?
A. 0.255 qt B. 270 qt C. 236 qt D. 0.270 qt E. 0.400 qt

17. An order for a patient calls for 0.020 grams of medication. On hand are 4-mg tablets. What dose is needed for the patient?
A. 2 tablets B. 4 tablets C. 5 tablets D. 8 tablets E. 10 tablets

18. A doctor orders 1500 mg of a sulfa drug. Tablets in stock are 0.500 g. How many tablets are needed?
A. 1 tablet B. 1½ tablets C. 2 tablets D. 2½ tablets E. 3 tablets

19. What is the density of a bone with a mass of 192 g and a volume of 120 cm^3?
A. 0.63 g/mL B. 1.4 g/cm^3 C. 1.6 g/mL
D. 1.9 g/cm^3 E. 2.8 g/cm^3

20. How many milliliters of a salt solution with a density of 1.8 g/mL are needed to provide 400 g of salt solution?
A. 220 mL B. 22 mL C. 720 mL D. 400 mL E. 4.5 mL

21. The density of a solution is 0.85 g/mL. Its specific gravity is
A. 222 mL B. 8.5 C. 0.85 mL D. 1.2 E. 0.85

22. Three liquids have densities of 1.15 g/mL, 0.79 g/mL and 0.95 g/mL. When the liquids, which do not mix, are poured into a graduated cylinder, the liquid at the top is the one with a density of

A. 1.15 g/mL B. 1.00 g/mL C. 0.95 g/mL D. 0.79 g/mL E. 0.16 g/mL

23. A sample of oil has a mass of 65 g and a volume of 80.0 mL. What is the specific gravity of the oil?
A. 1.5 B. 1.4 C. 1.2 D. 0.90 E. 0.81

24. What is the mass of a 10.0 mL sample of urine with a specific gravity of 1.04?
A. 104 g B. 10.4 g C. 1.04 g D. 1.40 g E. 0.140 g

25. Ethyl alcohol has a density of 0.790 g/mL. What is the mass of 0.250 L of the alcohol?
A. 198 g B. 158 g C. 3.95 g D. 0.253 g E. 0.160 g

ANSWERS FOR THE SELF-EVALUATION TEST

1. C	6. B	11. E	16. D	21. E
2. D	7. B	12. A	17. C	22. D
3. B	8. E	13. D	18. E	23. E
4. D	9. D	14. D	19. C	24. B
5. A	10. A	15. B	20. A	25. A

SCORING THE SELF-EVALUATION TEST

questions 1-25 4 points each 100 points total

CHAPTER 2

ENERGY AND MATTER

KEY CONCEPTS

1. In the metric system, the temperature of a substance is stated in degrees Celsius or Kelvins.

2. Energy is defined as the ability to do work.

3. Heat energy may be measured in calories: one calorie of heat raises the temperature of one gram of water by one degree Celsius.

4. Matter is described as a solid, liquid, or gas. A solid has a definite shape and volume; a liquid changes shape, but has a definite volume; a gas changes both its shape and volume.

5. A change of state such as melting or boiling appears as a plateau on a heating curve.

6. A change of state requires or releases a specific amount of heat (kcal, calories or joules) per gram of substance.

7. The caloric value of a foodstuff represents the energy released by the combustion of that food (kcal/g).

KEY WORDS

Using complete sentences, write a description of the following key words:

calorie

specific heat

states of matter

heat of fusion

LEARNING EXERCISES

TEMPERATURE MEASUREMENT (2.1)

Review: The equation °F = 1.8°C + 32 is used to convert a Celsius temperature to a Fahrenheit temperature. When it is rearranged for °C, it may be used to convert from °F to °C.

$$°C = \frac{(°F - 32)}{1.8}$$

The temperature on the Celsius scale is related to the Kelvin scale as K = °C + 273.

A. Calculate the temperatures in the following problems:

1. At a local yogurt shop, the milk is warmed to 68°C in the preparation of yogurt. What Fahrenheit temperature is needed to prepare the yogurt?

2. On a cold day in Alaska, the temperature drops to −12°C. What is that temperature on a Fahrenheit thermometer?

3. A patient has a temperature of 39.5°C. What is that temperature in °F?

4. On a hot summer day, the temperature is 95°F. What is the temperature on the Celsius scale?

5. A pizza is cooked at a temperature of 425°F. What is the °C temperature?

6. A salt melts at 815°C. What is the temperature in Kelvins?

7. A research experiment requires the use of liquid nitrogen to cool the reaction flask to −45°C. What temperature will this be on the Kelvin scale?

8. What is a temperature of 85°F on the Kelvin scale?

Answers: (1) 154°F (2) 10°F (3) 103.1°F (4) 35°C (5) 218°C
(6) 1088 K (7) 228 K (8) 302 K

ENERGY (2.2)

B. Match the words in column A with the descriptions in column B

A	B
1._____kinetic energy	a. The heat required to raise the temperature of 1 g of water 1°C
2._____calorie	b. 1000 calories
3._____potential energy	c. The ability to do work.
4._____chemical energy	d. The energy of motion
5._____energy	e. A measure of the ability of a substance to absorb heat
6._____specific heat	f. Inactive or stored energy
7._____kilocalorie	g. The energy available in the bonds of chemical compounds

Answers: 1. d 2. a 3. f 4. g 5. c 6. e 7. b

C. State whether the following statements describe potential (P) or kinetic (K) energy:

_____1. A potted plant sitting on a window ledge

_____2. Your breakfast cereal

_____3. Logs sitting in a fireplace

_____4. Intravenous glucose solution in the bottle

_____5. An arrow shot from a bow

_____6. A ski jumper at the top of the ski jump

_____7. A jogger running

_____8. A sky diver waiting to jump

_____9. Water flowing down a stream

_____10. A bowling ball striking the pins

Answers: 1. P 2. P 3. P 4. P 5. K 6. P 7. K 8. P 9. K 10. K

MEASURING HEAT ENERGY (2.3)

Review: ***Specific heat*** is the amount of heat required to increase the temperature of 1 g of a substance by 1°C. For liquid water, it is defined as 1.00 calorie/g°C or 4.18 joules/g°C. To calculate the calories lost or gained by a substance during a change in temperature, we use

heat (calories) = mass (g) x temperature change x specific heat

D. Calculate the calories gained or released during the following:

1. Heating 20.0 g of water from 20.0°C to 75.0°C.

2. Heating 10.0 g of water from 10.0°C to 95.0°C.

3. Cooling 4.00 kg of water from 80.0°C to 35.0°C.

4. Cooling 5.0 g of water from 40.0°C to 60.0°C.

Answers: (1) 1100 cal (2) 850 cal (3) 180,000 cal (4) 100 cal

STATES OF MATTER (2.4)

E. State whether the following statements describe a gas(G), a liquid(L), or a solid(S).

1._____There are no attractions among the molecules.

2._____Particles are held close together in a definite pattern.

3._____The substance has a definite volume, but no definite shape.

4._____The particles are moving extremely fast.

5._____This substance has no definite shape and no definite volume.

6._____The particles are very far apart.

7._____This material has its own volume, but takes the shape of its container.

8._____The particles of this material are bombarding the sides of the container with great force.

9._____The particles in this substance are moving very, very slowly.

10._____This substance has a definite volume and a definite shape.

Answers: 1. G 2. S 3. L 4. G 5. G 6. G 7. L 8. G 9. S 10. S

CHANGES OF STATE (2.5)

F. Draw heating or cooling curves for the following substances. Indicate on each the portion of the curve that corresponds to a solid, liquid, gas, and the changes in state.

1. Draw a heating curve for water that begins at −20°C and ends at 120°C. Water has a melting point of 0°C and a boiling point of 100°C.

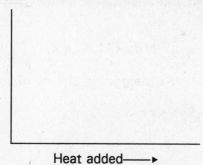

Heat added⟶

2. Draw a heating curve for bromine from −25° to 75°C. Br_2 has a melting point of −7°C and a boiling point of 59°C.

Heat added⟶

3. Draw a cooling curve for sodium from 1000°C to 0°C. Sodium has a freezing point of 98°C and a boiling (condensation) point of 883°C.

Heat removed⟶

Answers:

1.

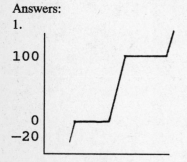

100

0

−20

2.

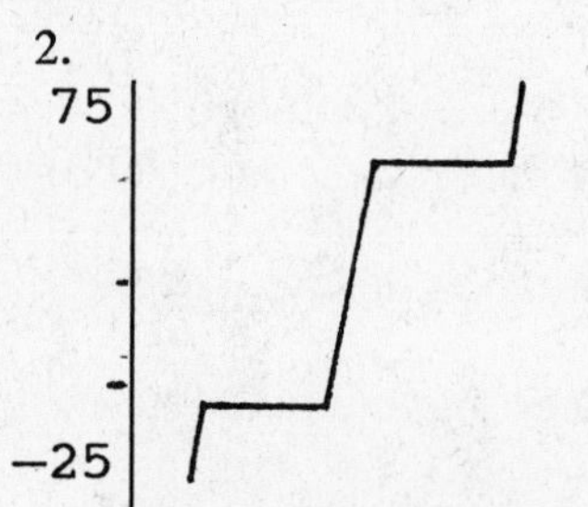

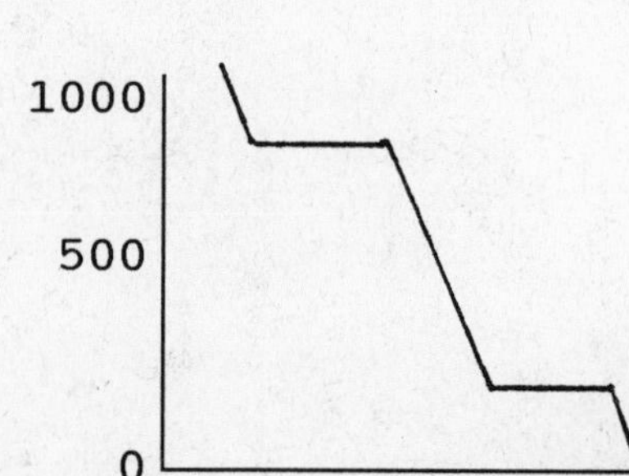

ENERGY IN CHANGES OF STATE (2.6)

Review: When water changes state from solid to liquid at 0°C, 80 calories, the heat of fusion, is the heat required to melt 1 g of solid water (ice); it is also the heat lost when 1 gram of water freezes at 0°C. When water boils at 100°C, 540 calories, the heat of vaporization, is required to change 1 g of liquid to gas (steam); it is also the amount of heat released when 1 g of water vapor condenses at 100°C. To calculate the amount of heat needed or released during a change of state, we use

Heat (cal) = mass (g) x heat of change of state

G. Calculate the energy for the following substances undergoing a change of state:

1. How many calories are needed to melt 15 g ice at 0°C?

2. How many calories are needed to vaporize 10.0 g H_2O at 100°C?

3. How many calories are needed to melt 15.0 g ice at 0°C, heat the water to 100°C, and convert the water to gas at 100°C?

4. How many kcal are released when 250 g H_2O at 65°C are cooled to 0°C and frozen?

5. How many kcal are absorbed when 150 g ice in an ice bag melt at 0°C and the water warms to body temperature of 37°C?

6. Calculate the number of kcal released when 50.0 g of steam at 100°C are cooled and frozen at 0°C.

Answers: (1) 1200 cal (2) 5400 cal (3) 10,800 cal (4) 36 kcal (5) 18 kcal (6) 36 kcal

CALORIC CONTENT OF FOOD (2.7)

H. State the caloric value in kcal/g associated with the following:

1. amino acid ________ 2. protein ________
3. sugar ________ 4. sucrose ________
5. oil ________ 6. fat ________
7. starch ________ 8. lipid ________
9. glucose ________ 10. lard ________

Answers: (1) 4 kcal/g (2) 4 kcal/g (3) 4 kcal/g (4) 4 kcal/g (5) 9 kcal/g (6) 9 kcal/g (7) 4 kcal/g (8) 9 kcal/g (9) 4 kcal/g (10) 9 kcal/g

I. Calculate the kcal for the following foods using the following calorimetry data:

Food	*Carbohydrate*	*Fat*	*Protein*	*kcal*
1. Peas, green, cooked	19 g	1 g	9 g	________
2. Potato chips, 10 chips	10 g	8 g	1 g	________
3. Cream cheese, 8-oz pkg	5 g	86g	18 g	________
4. Hamburger, lean, 3 oz	0	10 g	23 g	________
5. Salmon, canned	0	5 g	17 g	________
6. Snap beans, 1 cup	7 g	2 g	30 g	________
7. Banana, 1	26 g	0	1 g	________

Answers: (1) 121 kcal (2) 116 kcal (3) 866 kcal (4) 182 kcal (5) 113 kcal (6) 166 kcal (7) 108 kcal

J. Using caloric values, calculate each of the following:

1. How many kcal are in a single serving of pudding that contains 4 g protein, 31 g of carbohydrate, and 5 g of fat.

2. A can of tuna has a caloric value of 200 kcal. If there are 2 g of fat and no carbohydrate, how many grams of protein are contained in the can of tuna?

3. A serving of breakfast cereal provides 220 kcal. In this serving, there are 8 g of protein and 6 g of fat.

   a) How many grams of carbohydrates are in the cereal?

   b) What percent of the calories is obtained from the carbohydrate in the cereal?

4. Complete the following table listing ingredients for a peanut butter sandwich.

	Protein	***Carbohydrate***	***Fat***	***kcal***
2 slices bread	4 g	30 g	0	______
2 Tb peanut butter	7.8 g	6.4 g	______	172 kcal
2 tsp jelly	0	______	0	40 kcal
1 tsp margarine	0	0	5g	______
	Total kcal in sandwich			______

Answers:
1. protein 16 kcal + carbohydrate 124 kcal + fat 45 kcal = 185 kcal
2. fat 18 kcal; 200 kcal − 18 = protein 182 kcal; 46 g protein
3. protein 32 kcal + fat 54 kcal = 86 kcal protein and fat;
   220 kcal − 86 kcal = 134 kcal due to carbohydrate;
   % calories from carbohydrate = 134 kcal/220 kcal x 100% = 60.9%
4. bread, 136 kcal; peanut butter, 12.8g fat; jelly, 10 g carbohydrate; margarine, 45 kcal;
   total kcal = 393 kcal

SELF–EVALUATION TEST

ENERGY AND MATTER

For questions 1,2,3, and 4, consider the heating curve below for p-toluidine. Answer the following questions when heat is added to p-toluidine at 20°C where toluidine is below its freezing point.

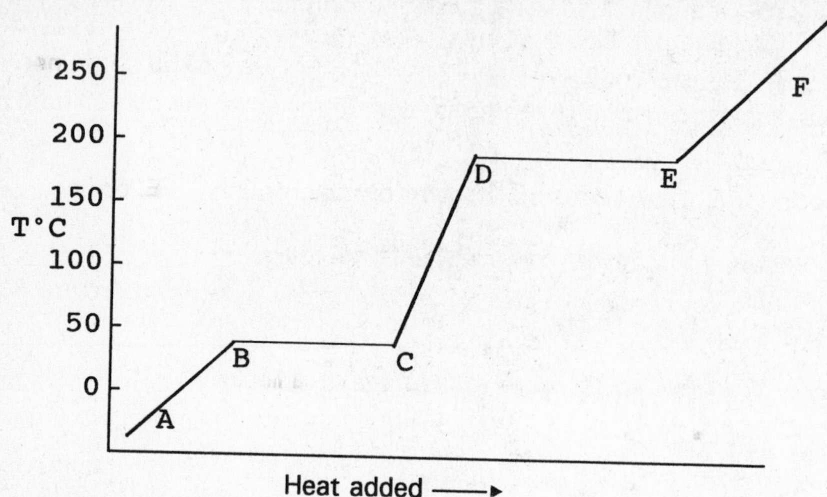

1. On the heating curve, segment BC indicates
   A. solid B. melting C. a liquid D. boiling E. a gas

2. On the heating curve, segment CD shows toluidine as
   A. solid B. melting C. a liquid D. boiling E. a gas

3. The boiling point of toluidine would be
   A. 20°C B. 45°C C. 100°C D. 200°C E. 250°C

4. On the heating curve, segment EF shows toluidine as
   A. solid B. melting C. a liquid D. boiling E. a gas

5. 105°F = ____ °C
   A. 73°C B. 41°C C. 58°C D. 90°C E. 189°C

6. The melting point of gold is 1064°C. The Fahrenheit temperature needed to melt gold would be
   A. 129°C B. 623°F C. 1031°F D. 1913°F E. 1947°F

7. The average daytime temperature on the planet Mercury is 683 K. What is this temperature on the Celsius scale?
   A. 956°C B. 715°C C. 680°C D. 410°C E. 303°C

8. Which of the following would be described as potential energy?
   A. A car going around a racetrack.
   B. A rabbit hopping.
   C. A deposit of undiscovered oil.
   D. A moving merry-go-round.
   E. A bouncing ball.

9. The number of calories needed to raise the temperature of 5.0 g water from 25°C to 55°C is
A. 5 cal B. 30 cal C. 5 cal D. 80 cal E. 150 cal

10. Which of the following describes a liquid?
A. A substance that has no definite shape and no definite volume.
B. A substance with particles that are far apart.
C. A substance with a definite shape and a definite volume.
D. A substance containing particles that are moving very fast.
E. A substance that has a definite volume, but takes the shape of its container.

Identify the following statements as:
A. energy B. evaporation C. heat of fusion D. heat of vaporization E. boiling

11. ____The energy required to convert a gram of solid to liquid.

12. ____The heat needed to boil a liquid.

13. ____The escape of liquid molecules from the surface of a liquid.

14. ____The ability to do work.

15. ____The formation of a gas within the liquid as well as on the surface.

16. Ice cools down a drink because
A. the ice is colder than the drink and heat flows into the ice cubes.
B. heat is absorbed from the drink to melt the ice cube.
C. the heat of fusion of the ice is higher than the heat of fusion for water.
D. Both A and B
E. None of the above

17. A can full of steam is tightly stoppered. As the can cools
A. nothing will happen.
B. the steam will blow up the can.
C. the steam condenses and the can collapses.
D. the steam condenses and the water formed will expand and explode the can.
E. None of the above

18. The number of kilocalories needed to convert 400 g of ice to liquid at 0°C is
A. 400 kcal B. 320 kcal C. 80 kcal D. 40 kcal E. 32 kcal

19. The number of calories released when 2 g water at 50°C are cooled and frozen at 0°C is
A. 100 cal B. 160 cal C. 260 cal D. 500 cal E. 820 cal

20. What is the total number of calories required to convert 25 g of ice at 0°C to gas at 100°C?
A. 2000 cal B. 4500 cal C. 14,000 D. 16,000 cal E.18,000 cal

21. The kcal needed to convert 10 g of ice at 0°C to steam at 100°C are
A. 0.8 kcal B. 1.8 kcal C. 6.2 kcal D. 6.4 kcal E. 7.2 kcal

For the following questions, consider a cup of milk that is 3.5% butterfat with a caloric value of 165 kcal. In the cup of milk, there are 9 g of fat, 12 g of carbohydrate, and some protein.

22. The number of kcal provided by the carbohydrate is
A. 4 kcal B. 9 kcal C. 36 kcal D. 48 kcal E. 81 kcal

23. The number of kcal provided by the fat is
A. 4 kcal B. 9 kcal C. 36 kcal D. 48 kcal E. 81 kcal

24. The number of kcal provided by the protein is
A. 4 kcal B. 9 kcal C. 36 kcal D. 48 kcal E. 81 kcal

25. The number of grams of protein provided in the cup of milk is
A. 4 g B. 9 g C. 36 g D. 48 g E. 81 g

ANSWERS TO THE SELF—EVALUATION TEST

1. B	6. E	11. C	16. D	21. E
2. C	7. D	12. D	17. C	22. D
3. D	8. C	13. B	18. E	23. E
4. E	9. E	14. A	19. C	24. C
5. B	10. E	15. E	20. E	25. B

SCORING THE SELF—EVALUATION TEST

25 questions 4 points each 100 points total

CHAPTER 3
ATOMS AND ELEMENTS

KEY CONCEPTS

1. Symbols are used to represent the names of elements.

2. An atom contains a nucleus with positively charged protons and neutral neutrons surrounded by a great amount of spaçe containing negatively charged electrons. A *neutral* atom contains an equal number of protons and electrons and has a charge of zero (0).

3. The atomic number of an element represents the number of protons in each atom of that element; the mass number represents the total number of protons and neutrons. Isotopes are atoms having the same atomic number, but different mass numbers.

4. Elements are located in the periodic table in a period and in a family.

5. The electrons in an atom are arranged in energy levels or shells beginning with the lowest energy and progressing to higher energy levels.

6. The energy levels or shells contain sublevels or subshells of closely related energies.

KEY WORDS

Use complete sentences to describe the following terms:

element

atom

atomic number

mass number = proton + neutrons

isotope

electron arrangement

LEARNING EXERCISES

ELEMENTS AND SYMBOLS (3.1)

A. Give the symbols for each of the following elements:

1.	carbon	______	2.	iron	______
3.	sodium	______	4.	phosphorus	______
5.	oxygen	______	6.	nitrogen	______
7.	iodine	______	8.	sulfur	______
9.	boron	______	10.	lead	______
11.	calcium	______	12.	gold	______
13.	copper	______	14.	neon	______
15.	potassium	______	16.	silicon	______

Answers: 1. C 2. Fe 3. Na 4. P 5. O 6. N 7. I 8. S
9. B 10. Pb 11. Ca 12. Au 13. Cu 14. Ne 15. K 16. Si

B. Write the names of the elements represented by each of the following symbols:

1. Mg ______________	2. K ______________
3. H ______________	4. F ______________
5. Cu ______________	6. Be ______________
7. Ag ______________	8. Br ______________
9. Zn ______________	10. Cl ______________
11. Ba ______________	12. Li ______________
13. Cl ______________	14. Al ______________
15. He ______________	16. Ni ______________

Answers: 1. magnesium 2. potassium 3. hydrogen 4. fluorine 5. copper 6. beryllium
7. silver 8. bromine 9. zinc 10. mercury 11. barium 12. lithium
13. chlorine 14. aluminum 15. helium 16. nickel

THE ATOM (3.2)

Review: An atom contains subatomic particles; protons (1 +), neutrons (0), and electrons (1-). The protons and neutrons are found in the *nucleus*, a dense, compact cluster in the center of the atom. The electrons occupy the space outside the nucleus.

C. Give the symbol, charge and location in the atom for the subatomic particles:

	Particle	*Symbol*	*Electrical charge*	*Location*
1.	Proton	_____	_____	_______________
2.	Neutron	_____	_____	_______________
3.	Electron	_____	_____	_______________

Answers: 1. p^+, 1+, nucleus; 2. n, 0, nucleus 3. e^-, 1-, outside, surrounding the nucleus

ATOMIC NUMBER AND MASS NUMBER (3.3)

Review: The *atomic number* is equal to the number of protons in the atom. The *mass number* is the sum of the number of protons and neutrons. *Isotopes* are atoms of the same element (same atomic number), but contain different numbers of neutrons.

D. Give the number of protons in a neutral atom

1. Of carbon. _______
2. With atomic number 15. _______
3. With a mass number of 40 and atomic number 19. _______
4. With 9 neutrons and a mass number of 19. _______
5. That has 18 electrons. _______

Answers: (1) 6 (2) 15 (3) 19 (4) 10 (5) 18

E. Find the number of neutrons in an atom with

1. A mass number of 42 and atomic number 20. _______
2. A mass number of 10 and 5 protons. _______
3. A mass number of 18 and 8 electrons. _______
4. A mass number of 9 and atomic number 4. _______
5. A mass number of 22 and 10 protons. _______

Answers: (1) 22 (2) 5 (3) 10 (4) 5 (5) 12

F. Complete the following table for neutral atoms.

element	*atomic number*	*mass number*	*number of protons*	*number of neutrons*	*number of electrons*
______	12	______	______	12	______
______	______	______	20	22	______
iron	______	55	______	______	______
______	26	______	______	31	______
______	______	35	17	______	______
______	______	21	______	______	10

Answers:

element	atomic number	mass number	number of protons	number of neutrons	number of electrons
magnesium	12	24	12	12	12
calcium	20	42	20	22	20
iron	26	55	26	29	26
iron	26	57	26	31	26
chlorine	17	35	17	18	17
neon	10	21	10	11	10

G. Identify the atoms that are isotopes.

A. $^{20}_{10}X$ B. $^{20}_{11}X$ C. $^{21}_{11}X$ D. $^{19}_{10}X$ E. $^{18}_{9}X$

Answers: Atoms A and D are isotopes; atoms B and C are isotopes.

H. ***Writing Exercise***: Copper has two naturally occurring isotopes, ^{63}Cu and ^{65}Cu. If that is the case, why is the atomic weight of copper listed in the periodic table as 63.5?

Answer: A sample of copper occurring in nature consists of two isotopes with different atomic masses. The atomic weight is the weighted average of these two isotopes and does not represent individual isotopes.

THE PERIODIC TABLE (3.4)

I. Indicate whether the following elements are in a family (F), period (P), or neither (N):

1. Li, C and O __________ 2. Br, Cl and F __________

3. Al, Si and Cl __________ 4. C, N and O __________

5. Mg, Ca, and Ba __________ 6. C, S and Br __________

7. Li, Na and K __________ 8. K, Ca and Br __________

Answers 1. P 2. F 3. P 4. P 5. F 6. N 7. F 8. P

J. Identify each of the following elements as a metal (M) or nonmetal (NM):

1. Cl _____ 2. N _____ 3. Fe _____ 4. K_____

5. Al _____ 6. C _____ 7. Ca _____ 8. Zn_____

9. Ag _____ 10. Ca _____ 11. O _____ 12. Li_____

Answers: 1. NM 2. NM 3. M 4. M 5. M 6. NM
7. M 8. M 9. M 10. M 11. NM 12. M

ELECTRON ARRANGEMENT IN THE ATOM (3.5)

Review: The electrons in an atom are arranged in order of increasing energy levels. In the main energy shells, the order of electrons for the first twenty (20) elements is 2, 8, 8, 2.

K. Write the electron arrangement for the following elements:

element	*electron shell (n)* 1 2 3 4
1. beryllium	________________
2. potassium	________________
3. calcium	________________
4. sodium	________________
5. phosphorus	________________
6. nitrogen	________________
7. chlorine	________________
8. silicon	________________

Answers: (1) 2,2 (2) 2, 8, 8, 1 (3) 2, 8, 8, 2 (4) 2, 8, 1
(5) 2, 8, 5 (6) 2, 5 (7) 2, 8, 7 (8) 2, 8, 4

PERIODIC LAW (3.6)

Review: The ***A group numbers*** correspond to the number of valence electrons contained in the outer shell of an atom.

L. State the number of electrons in the outer electron level and the group number of each element.

element	*electrons in outer level*	*group number*
1. sulfur	__________	__________
2. oxygen	__________	__________
3. magnesium	__________	__________
4. hydrogen	__________	__________
5. fluorine	__________	__________
6. aluminum	__________	__________
7. boron	__________	__________
8. chlorine	__________	__________

Answers: (1) 6, 6A (2) 6, 6A (3) 2, 2A (4) 1, 1A
(5) 7, 7A (6) 3, 3A (7) 3, 3A (8) 7, 7A

SELF–EVALUATION TEST
ATOMS AND ELEMENTS

Write the correct symbol for each of the elements listed:

1. ______ potassium

2. ______ phosphorus

3. ______ calcium

4. ______ carbon

5. ______ sodium

Write the correct name for each of the symbols listed:

6. Fe ________________

7. Cu ________________

8. Cl ________________

9. Pb ________________

10. Ag ________________

For questions 11-14, consider an atom with 12 protons and 13 neutrons.

11. This atom has an atomic number of
A. 12 B. 13 C. 23 D. 24 E. 25

12. This atom has a mass number of
A. 12 B. 13 C. 23 D. 24.3 E. 25

13. This is an atom of
A. carbon B. sodium C. magnesium D. silicon E. manganese

14. The number of electrons in this atom is
A. 12 B. 13 C. 23 D. 24 E. 25

For questions 15-18, consider an atom of calcium with a mass number of 42.

15. This atom of calcium has an atomic number of
A. 20 B. 22 C. 40 D. 41 E. 42

16. The number of protons in this atom of calcium is
A. 20 B. 22 C. 40 D. 41 E. 42

17. The number of neutrons in this atom of calcium is
A. 20 B. 22 C. 40 D. 41 E. 42

18. The number of electrons in an atom of calcium with mass number 42 is
A. 20 B. 22 C. 40 D. 41 E. 42

19. Platinum, ^{195}Pt, has
A. 78 p^+, 78e^-, 78n B. 195 p^+, 195e^-, 195n C. 78p^+, 78e^-, 117n
D. 78p^+, 78e^-, 117n E. 78p^+, 117e^-, 117n

For questions 20-21, use the following list of atoms.

$^{14}_{7}V$ $^{16}_{8}W$ $^{19}_{9}X$ $^{16}_{7}Y$ $^{18}_{8}Z$

20. Which atoms(s) are isotopes of an atom with 8 protons and 9 neutrons?
A. W B. W, Z C. X, Y D. X E. Y

21. Which atom(s) are isotopes of an atom with 7 protons and 8 neutrons?
A. V B. W C. V, Y D. W, Z E. none

22. Which element would you expect to have properties most like oxygen?
A. nitrogen B. carbon C. chlorine D. argon E. sulfur

23. Which of the following is an isotope of nitrogen?

A. $^{14}_{8}N$ B. $^{7}_{3}N$ C. $^{10}_{5}N$ D. $^{4}_{2}N$ E. $^{15}_{7}N$

24. The elements C, N, and O are part of a
A. period B. family C. neither

25. The element Li, Na, and K are part of a
A. period B. family C. neither

26. What is the classification of an atom with 15 protons and 17 neutrons?
A. metal B. nonmetal C. transition element D. noble gas E. halogen

27. What is the group number of the element with atomic number 3?
A. 1A B. 2A C. 3A D. 7A E. 8A

28. After helium, the number of electrons in the outer shells of the noble gases.
A. 3 B. 5 C. 7 D. 8 E. 12

29. The electron configuration for an oxygen atom is
A. 2,8 B. 2,2,4 C. 2,6,8 D. 8,2,6 E. 2,6

30. The electron configuration for aluminum is
A. 2,8,2,1 B. 2,6,5 C. 2,8,3 D. 2,2,8,1 E. 10,3

For questions 31-36, select answers from the following:
A. 2 B. 3 C. 4 D. 6 E. 8

31. The number of electrons in the second energy level of phosphorus.

32. The number of electrons in the first energy level of carbon

33. The number of electrons in the highest energy level of silicon.

34. The number of electrons in the second energy level of boron.

35. The number of electrons in the fourth energy level of calcium.

36. The number of electrons in the highest energy level of argon.

Electrons are arranged about the nucleus of an atom in higher and higher energy levels. For questions 37-40, indicate whether the conditions will cause an electron to jump to a higher energy level, fall to a lower energy level, or show no change.

A. electron jump B. electron fall C. no change

37. An atom in the ground state with no outside source of energy.

38. Heat is applied to an atom.

39. Color is seen in a flame test of an element.

40. Energy is released by an atom.

ANSWERS FOR THE SELF-EVALUATION TEST

1. K	6. iron	11. A	16. A	21. C	26. B	31. E	36. E
2. P	7. copper	12. E	17. B	22. E	27. A	32. A	37. C
3. Ca	8. chlorine	13. C	18. A	23. E	28. D	33. C	38. A
4. C	9. lead	14. A	19. D	24. A	29. E	34. B	39. B
5. Na	10. silver	15. A	20. B	25. B	30. C	35.A	40. B

SCORING THE SELF-EVALUATION TEST

40 questions 2½ points each 100 points total

CHAPTER 4
COMPOUNDS AND THEIR BONDS

KEY CONCEPTS

1. The electrons in the outer energy level of an atom (valence electrons) are responsible for much of the chemical activity of an element.

2. Ions are formed when atoms lose or gain electrons.

3. The charge on an ion represents the loss or gain of one or more electrons by an atom. Ionic bonding is the electrical attraction between oppositely charged ions.

4. Ionic formulas represent a balance of positive and negative ionic charges so that the overall charge of the formula is zero.

5. The names of ionic compounds indicate the metal ion first, followed by the name of the nonmetal ion.

6. Covalent bonding is the result of a sharing of electrons between two nonmetals.

7. Polar bonds in covalent compounds occur between nonmetals with unequal electronegativities. Nonpolar covalent bonds occur between nonmetals that have identical electronegativity values.

8. The names of covalent compounds with two nonmetals include prefixes to indicate any subscripts in the formula.

9. Polyatomic ions consist of groups of atoms that have a net ionic charge.

KEY WORDS

Using complete sentences, describe each of the following terms:

electron dot structure

ion

ionic bond

covalent bond

LEARNING EXERCISES

VALENCE ELECTRONS (4.1)

A. Add dots to the following symbols to give their electron dot structures:

1. O 2. Cl 3. Na 4. Ca 5. Li

6. H 7. P 8. S 9. Mg 10. C

Answers: 1. : Ö · (with one dot below) 2. : C̈l : (with one dot below) 3. Na · 4. Ċa· 5. Li ·

6. H · 7. · P̈ · (with one dot below) 8. : S̈ · (with one dot below) 9. Ṁg · 10. · Ċ · (with one dot below)

THE OCTET RULE (4.2)

B. For the following elements, write their electron arrangements, the number of valence electrons, an octet, and whether it is likely to form a compound.

Element	*Electron Arrangement*	*Number of valence electrons*	*Octet (yes/no)*	*Formation of a compound (yes/no)*
1. C	______	4	n	y
2. Ne	______	8	y	n
3. Ca	______	2	n	y
4. Ar	______	8	y	n
5. N	______	5	n	y

Answers:

1. C $2e^-$,$4e^-$; 4 valence electrons; no, yes

2. Ne $2e^-$ $8e^-$; 8 valence electrons; yes; no

3. Ca $2e^-$ $8e^-$ $8e^-$ $2e^-$; 2 valence electrons; no; yes

4. Ar $2e^-$ $8e^-$ $8e^-$; 8 valence electrons; yes; no

5. N $2e^-$ $5e^-$; 5 valence electrons; no; yes

IONS (4.3)

C. The following elements lose electrons when they form ions. Indicate the group number, the number of electrons lost and the ion (symbol and charge) for each of the following:

	Element	*Group number*	*Electrons lost*	*Ion formed*
1.	magnesium	2A	$2e^-$	Mg^{2+}
2.	sodium	1A	$1e^-$	Na^+
3.	calcium	2A	$2e^-$	Ca^{2+}
4.	potassium	1A	$1e^-$	K^+
5.	aluminum	3A	$3e^-$	Al^{3+}

Answers:
1. 2A, $2e^-$ lost, Mg^{2+}
2. 1A, $1e^-$ lost, Na^+
3. 2A, $2e^-$ lost, Ca^{2+}
4. 1A, $1e^-$ lost, K^+
5. 3A, $3e^-$ lost, Al^{3+}

D. The following elements gain electrons when they form ions. Indicate the group number, the number of electrons gained and the ion (symbol and charge) for each of the following:

	Element	*Group number*	*Electrons gained*	*Ion formed*
1.	chlorine	7A	$1e^-$	Cl^-
2.	oxygen	6A	$2e^-$	O^{2-}
3.	nitrogen	5A	$3e^-$	N^{3-}
4.	fluorine	7A	$1e^-$	F^-
5.	sulfur	6A	$2e^-$	S^{2-}

Answers:
1. 7A, gain $1e^-$, Cl^-
2. 6A, gas $2e^-$, O^{2-}
3. 5A, gain $3e^-$, N^{3-}
4. 7A, gain $1e^-$, F^-
5. 6A, gain $2e^-$, S^{2-}

E. Most of the transition metals form two or more ions with positive charge. Complete the table:

	Name of ion	*Symbol of ion*
1.	iron (III)	Fe^{3+}
2.	Copper (II)	Cu^{2+}
3.	zinc	Zn^{2+}
4.	iron (II)	Fe^{2+}
5.	copper (I)	Cu^{+}
6.	silver	Ag^{+}

Answers: 1. Fe^{3+} 2. copper (II) 3. Zn^{2+} 4. iron(II) 5. Cu^{+} 6. Ag^{+}

IONIC COMPOUNDS (4.4)

Review: In the formulas of ionic compounds, the total positive charge is equal to the total negative charge. When two or more ions are needed for charge balance, that number is indicated by subscripts in the formula. For example, the compound magnesium chloride, $MgCl_2$, contains Mg^{2+} and 2 Cl^-. The sum of the charges is zero: $(2+) + 2(-) = 0$.

F. Write the correct ionic formula for the compound formed from the following pairs of ions:

1. Na^+, Cl^- $NaCl$
2. Al^{3+}, O^{2-} Al_2O_3
3. Mg^{2+}, Cl^- $MgCl_2$
4. Fe^{2+}, F^- FeF_2
5. K^+, S^{2-} K_2S
6. Fe^{3+}, O^{2-} Fe_2O_3
7. Cu^+, O^{2-} Cu_2O
8. Al^{3+}, Cl^- $AlCl_3$

Answers: 1. $NaCl$ 2. Al_2O_3 3. $MgCl_2$ 4. FeF_2 5. K_2S 6. Fe_2O_3 7. Cu_2O 8. $AlCl_3$

G. Write the ions and the correct ionic formula for the following ionic compounds:

		Ions (+)	Ions (−)	Formula of compound
1.	aluminum sulfide	Al^{3+}	S^{2-}	Al_2S_3
2.	copper (II) chloride	Cu^{2+}	Cl^-	$CuCl_2$
3.	magnesium oxide			
4.	iron (II) bromide			
5.	silver oxide			

Answers
1. Al^{3+}, S^{2-}, Al_2S_3
2. Cu^{2+}, Cl^-, $CuCl_2$
3. Mg^{2+}, O^{2-}; MgO
4. Fe^{2+}, Br^-; $FeBr_2$
5. Ag^+, O^{2-}; Ag_2O

NAMING IONIC COMPOUNDS (4.5)

H. Write the names of the following ions:

1. Cl^- __________ 2. Fe^{2+} __________

3. Cu^+ __________ 4. Ag^+ __________

5. O^{2-} __________ 6. Ca^{2+} __________

7. S^{2-} __________ 8. Al^{3+} __________

9. Fe^{3+} __________ 10. Ba^{2+} __________

11. Cu^{2+} __________ 12. N^{3-} __________

Answers: 1. chloride 2. iron (II) 3. copper (I) 4. silver 5. oxide 6. calcium 7. sulfide 8. aluminum 9. iron (III) 10. barium 11. copper (II) 12. nitride

I. Write a correct name for the following ionic compounds:

1. $BaCl_2$ ______________________

2. $FeBr_3$ ______________________

3. Na_3P ______________________

4. Al_2O_3 ______________________

5. CuO ______________________

6. Mg_3N_2 ______________________

Answers: 1. barium chloride 2. iron (III) bromide 3. sodium phosphide

4. aluminum oxide 5. copper (II) oxide 6. magnesium nitride

COVALENT BONDS (4.6)

Review: In covalent compounds, electrons are shared in a covalent bond to achieve an octet. For example oxygen with six valence electrons shares electrons with two hydrogen atoms to form the covalent compound water (H_2O).

$$\begin{array}{c} \quad\;\,\cdot\cdot \\ H : \ddot{O} : \\ \quad\;\, H \end{array}$$

J. List the number of bonds typically formed by the following atoms in covalent compounds:

1. N 3   2. S 2   3. P 3   4. C 4

5. Cl 1   6. O 2   7. H 1   8. F 1

Answers: (1) 3 (2) 2 (3) 3 (4) 4 (5) 1 (6) 2 (7) 1 (8) 1

K. Write the electron dot structures for the following covalent compounds:

H_2 NCl_3 HCl

Cl_2 H_2S CCl_4

Answers: H : H   Cl : N : Cl (with Cl below N)   H : Cl   Cl : Cl   H : S (with H below S)   Cl : C : Cl (with Cl above and below C)

BOND POLARITY (4.7)

Review: When atoms sharing electrons have the same electronegativity values (usually atoms of the same element), the pair is shared equally and the bond is nonpolar. When the electronegativity difference is between 0 and 1.7, the bond is polar. If the difference is 1.7 or greater, the bond is considered ionic.

L. Identify the bonding between the following pairs of elements as ionic (I), polar (P), covalent(C), or none.

1. H and S _____
2. P and O _____
3. Mg and O _____
4. Li and F _____
5. H and Cl _____
6. Cl and Cl _____
7. S and F _____
8. He and He _____

Answers 1. P 2. P 3. I 4. I 5. P 6. C 7. P 8. none

NAMING COVALENT COMPOUNDS (4.8)

M. Use the appropriate prefixes in naming the following covalent compounds:

1. CS_2 ______
2. CCl_4 ______
3. CO ______
4. SO_2 ______
5. N_2O_4 ______
6. PCl_3 ______
7. N_2O ______
8. SF_6 ______

Answers: 1. carbon disulfide 2. carbon tetrachloride 3. carbon monoxide 4. sulfur dioxide 5. dinitrogen tetroxide 6. phosphorus trichloride 7. dinitrogen monoxide 8. sulfur hexafluoride

POLYATOMIC IONS (4.9)

N. Write the polyatomic ion (symbol and charge) for the following:

1. sulfate ion SO_4^{2-}
2. hydroxide ion OH^-
3. carbonate ion CO_3^{2-}
4. sulfite ion SO_3^{2-}
5. ammonium ion NH_4^+
6. phosphate ion PO_4^{3-}
7. nitrate ion NO_3^-
8. nitrite ion NO_2^-

Answers 1. SO_4^{2-} 2. OH^- 3. CO_3^{2-} 4. SO_3^{2-} 5. NH_4^+ 6. PO_4^{3-} 7. NO_3^- 8. NO_2^-

O. Write the formula of each ion or polyatomic ion, and the correct formula for the following compounds:

	Positive ion	*Negative ion*	*Formula*
1. sodium phosphate	______	______	______________
2. ferric hydroxide	______	______	______________
3. ammonium carbonate	______	______	______________
4. silver bicarbonate	______	______	______________
5. iron (III) sulfate	______	______	______________
6. ferrous nitrate	______	______	______________
7. potassium sulfite	______	______	______________
8. barium phosphate	______	______	______________

Answers: 1. Na^+, PO_4^{3-}, Na_3PO_4 2. Fe^{3+}, OH^-, $Fe(OH)_3$

3. NH_4^+, CO_3^{2-}, $(NH_4)_2CO_3$ 4. Ag^+, HCO_3^-, $AgHCO_3$

5. Fe^{3+}, SO_4^{2-}, $Fe_2(SO_4)_3$ 6. Fe^{2+}, NO_3^-, $Fe(NO_3)_2$

7. K^+, SO_3^{2-}, K_2SO_3 8. Ba^{2+}, PO_4^{3-}, $Ba_3(PO_4)_2$

SUMMARY OF WRITING FORMULAS AND NAMES

Review: In both ionic and covalent compounds of *two* different elements, the name of the element written first is the same as its name as an uncombined element. The name of the second element in a formula is given the ending "ide". For example, $BaCl_2$ is named *barium chloride*. If the metal forms positive ions with two or more possible ionic charges, a Roman numeral is added to its name to indicate the ionic charge in the compound. For example, $FeCl_3$ is named *iron (III) chloride*. In naming covalent compounds, a prefix between the name of the elements indicates the numerical value of any subscripts. For example, N_2O_3 is named *dinitrogen trioxide*.

In ionic compounds with three or more elements, there is a polyatomic ion present. Most are negative polyatomic ions with names that end in "ate" or "ite", except for hydroxide. For example, Na_2SO_4 is named *sodium sulfate*.

P. If the compound formed from the components is ionic, write the ions; if it is covalent, write the electron dot structure. Then give a formula and name for each.

	Components	*Bond type ionic (I) or covalent (C)*	*Ions or dot structure*	*Formula*	*Name*
1.	Mg and Cl				
2.	N and Cl				
3.	K and SO_4				
4.	Li and O				
5.	Al and S				
6.	C and Cl				
7.	Na and PO_4				
8.	H and S				
9.	Cl and Cl				
10.	Ca and HCO_3				

Answers:

1. ionic, Mg^{2+}, Cl^-, $MgCl_2$, magnesium chloride
2. covalent, $\cdot\ddot{N}\cdot$ $:\ddot{Cl}\cdot$, NCl_3, nitrogen trichloride
3. ionic, K^+, SO_4^{2-}, K_2SO_4, potassium sulfate
4. ionic, Li^+, O^{2-}, Li_2O, lithium oxide
5. ionic, Al^{3+}, S^{2-}, Al_2S_3, aluminum sulfide
6. covalent, $\cdot\dot{C}\cdot$ $:\ddot{Cl}\cdot$, CCl_4, carbon tetrachloride
7. ionic, Na^+, PO_4^{3-}, Na_3PO_4, sodium phosphate
8. covalent, $H\cdot$ $:\ddot{S}\cdot$, H_2S, dihydrogen sulfide
9. covalent, $:\ddot{Cl}\cdot$ $:\ddot{Cl}\cdot$, Cl_2, chlorine
10. ionic, Ca^{2+}, HCO_3^-, $Ca(HCO_3)_2$, calcium hydrogen carbonate or calcium bicarbonate

SELF-EVALUATION TEST

COMPOUNDS AND THEIR BONDS

For questions 1-4, consider an atom of phosphorus.

1. It is in group
   A. 2A B. 3A C. 5A D. 7A E. 8A

2. How many valence electrons does it have?
   A. 2 B. 3 C. 5 D. 8 E. 15

3. To achieve an octet, the phosphorus atom will
   A. lose 1 electron B. lose 2 electrons C. lose 5 electrons
   D. gain 2 electrons E. gain 3 electrons

4. As an ion, it has an ionic charge (valence) of
   A. 1+ B. 2+ C. 5+ D. 2− E. 3−

5. To achieve an octet, a calcium atom
   A. loses 1 electron B. loses 2 electrons C. loses 3 electrons
   D. gains 1 electron E. gains 2 electrons

6. To achieve an octet, a chlorine atom
   A. loses 1 electron B. loses 2 electrons C. loses 3 electrons
   D. gains 1 electron E. gains 2 electrons

7. Another name for a positive ion is
   A. anion B. cation C. protion D. proton E. sodium

8. The correct ionic charge (valence) for calcium ion is
   A. 1+ B. 2+ C. 1− D. 2− E. 3−

9. The cation of silver has an ionic charge of
   A. 1+ B. 2+ C. 1− D. 2− E. 3−

10. The correct ionic charge (valence) for phosphate ion is
    A. 1+ B. 2+ C. 1− D. 2− E. 3−

11. The correct ionic charge (valence) for fluoride is
    A. 1+ B. 2+ C. 1− D. 2− E. 3−

12. The correct ionic charge (valence) for sulfate ion is
    A. 1+ B. 2+ C. 1− D. 2− E. 3−

13. When the elements magnesium and sulfur are mixed,
    A. an ionic compound forms.
    B. a covalent compound forms.
    C. no reaction occurs.
    D. the two repel each other and won't combine.
    E. None of the above.

14. An ionic bond typically occurs between
    A. two different nonmetals

14. An ionic bond typically occurs between
A. two different nonmetals
B. two of the same type of nonmetals
C. two noble gases
D. two different metals
E. a metal and a nonmetal

15. A covalent bond typically occurs between
A. two different nonmetals.
B. two of the same type of nonmetals.
C. two noble gases.
D. two different metals.
E. a metal and a nonmetal.

16. A polar bond typically occurs between
A. two different nonmetals.
B. two of the same type of nonmetals.
C. two noble gases.
D. two different metals.
E. a metal and a nonmetal.

17. The formula for a compound between carbon and chlorine is
A. CCl B. CCl_2 C. C_4Cl D. CCl_4 E. C_4Cl_2

18. The formula for a compound between sodium and sulfur is
A. SoS B. NaS C. Na_2S D. NaS_2 E. Na_2SO_4

19. The formula for a compound between aluminum and oxygen is
A. AlO B. Al_2O C. AlO_3 D. Al_2O_3 E. Al_3O_2

20. The formula for a compound between barium and sulfur is
A. BaS B. Ba_2S C. BaS_2 D. Ba_2S_2 E. $BaSO_4$

21. The correct formula for iron (III) chloride is
A. FeCl B. $FeCl_2$ C. Fe_2Cl D. Fe_3Cl E. $FeCl_3$

22. The correct formula for ammonium sulfate is
A. AmS B. $AmSO_4$ C. $(NH_4)_2S$ D. NH_4SO_4 E. $(NH_4)_2SO_4$

23. The correct formula for copper (II) chloride is
A. CoCl B. CuCl C. $CoCl_2$ D. $CuCl_2$ E. Cu_2Cl

24. The correct formula for lithium phosphate is
A. $LiPO_4$ B. Li_2PO_4 C. Li_3PO_4 D. $Li_2(PO_4)_3$ E. $Li_3(PO_4)_2$

25. The correct formula for silver oxide is
A. AgO B. Ag_2O C. Ag_2O D. Ag_3O_2 E. Ag_3O

26. The correct formula for magnesium carbonate is
A. $MgCO_3$ B. Mg_2CO_3 C. $Mg(CO_3)_2$ D. MgCO D. $Mg_2(CO_3)_3$

27. The correct formula for copper (I) sulfate is
A. $CuSO_3$ B. $CuSO_4$ C. Cu_2SO_3 D. $Cu(SO_4)_2$ E. Cu_2SO_4

28. The name of $AlPO_4$ is
A. aluminum phosphide B. alum phosphate C. aluminum phosphate
D. aluminum phosphorus oxide E. aluminum phosphite

29. The name of CuS is
A. copper sulfide B. copper (I) sulfate C. copper (I) sulfide
D. cuprous sulfide E. copper (II) sulfide

30. The name of $FeCl_2$ is
A. ferric chloride B. ferrous chloride C. iron (II) chlorine
D. iron chloride E. iron (III) chloride

31. The name of $ZnCO_3$ is
A. zinc(III) carbonate B. zinc (II) carbonate C. zinc bicarbonate
D. zinc carbon trioxide E. zinc carbonate

32. The name of Al_2O_3 is
A. aluminum oxide B. aluminum (II) oxide C. aluminum trioxide
D. dialuminum trioxide E. aluminum oxygenate

33. The name of NCl_3 is
A. nitrogen chloride B. nitrogen trichloride C. nitrogen (III) chloride
D. nitrogen chlorine three E. nitrogen chloride (III)

34. The name of CO is
A. carbon monoxide B. carbonic oxide C. carbon oxide
D. carbonious oxide E. carboxide

For questions 35-40, indicate the type of bonding expected between the following elements: A. ionic B. covalent C. polar D. none

35. ____silicon and oxygen

36. ____barium and chlorine

37. ____aluminum and chlorine

38. ____chlorine and chlorine

39. ____sulfur and oxygen

40. ____neon and oxygen

ANSWERS TO THE SELF-EVALUATION TEST

1.C	6.D	11.C	16.A	21.E	26.A	31.E	36.A
2.C	7.B	12.D	17.D	22.E	27.E	32.A	37.A
3.E	8.B	13.A	18.C	23.D	28.C	33.B	38.B
4.E	9.A	14.E	19.D	24.C	29.E	34.A	39.C
5.B	10.E	15.B	20.A	25.B	30.B	35.C	40.D

SCORING THE SELF-EVALUATION TEST

40 questions 2½ points each 100 points total

CHAPTER 5
CHEMICAL QUANTITIES

KEY CONCEPTS

1. The formula weight of a compound is the sum of the weights contributed by each element in the compound.
2. A mole is a quantity containing a large number (6.02×10^{23}) of particles.
3. One mole of an element or a compound has a mass in grams that is numerically equal to its atomic or formula mass.
4. A substance undergoes a physical change when the initial substance is retained. In a chemical change, the substance undergoes a reaction that gives new substances.
5. A chemical equation uses chemical symbols to describe the chemical changes that occur in a chemical reaction.
6. A chemical equation is balanced by placing coefficients in front of appropriate formulas to equalize the number of atoms of each element in the reactants and products.
7. A balanced chemical equation is used to calculate the quantities of products or reactants in a reaction.

KEY WORDS

Use complete sentences to write a description for each of the following key words:

formula weight

mole

molar mass

chemical change

chemical equation

LEARNING EXERCISES

A. FORMULA WEIGHT (5.1)

Review: Steps in Formula Weight Determination

1. Determine the number of atoms (subscripts) of each element in the compound.
2. Obtain the atomic weight of each element from the periodic table.
3. Calculate the total weight contributed by each element to the total weight of the compound.
4. Find the sum of all the weights of the elements to obtain the formula weight.

Example: What is the formula weight of silver nitrate, $AgNO_3$?

Solution:

	Step 1		Step 2		Step 3
Elements	Number of atoms	X	Atomic weight	=	Formula weight
Ag	1	X	107.9 amu	=	107.9 amu
N	1	X	14.0 amu	=	14.0 amu
O	3	X	16.0 amu	=	48.0 amu
	Step 4: Formula weight			=	169.9 amu

A. Determine the formula weight for the following compounds:

1. K_2O
2. $AlCl_3$
3. $Mg(NO_3)_2$
4. C_4H_{10}
5. O_2
6. Mg_3N_2
7. $FeCO_3$
8. $(NH_4)_3PO_4$

Answers: (1) 94.2 amu (2) 133.5 amu (3) 148.3 amu (4) 58.0 amu (5) 32.0 amu (6) 100.9 amu (7) 115.8 amu (8) 149.0 amu

THE MOLE (5.2)

Review: One mole of any substance contains the same number (Avogadro's number) of particles, 6.02×10^{23} and has a molar mass in grams equal to its formula weight. For example, HCl has a formula weight of 36.5 amu; one mole of HCl has a molar mass of 36.5 g.

B. Calculate the molar mass (g/mole) for the following:

1. Na_3PO_4
2. CO_2
3. Cl
4. $BaSO_4$
5. $Al_2(CO_3)_3$
6. C_5H_{12}

Answers: (1) 164.0 g/mole (2) 44.0 g/mole (3) 35.5 g/mole
(4) 233.4 g/mole (5) 234.0 g/mole (6) 72.0 g/mole

CALCULATIONS USING MOLAR MASS (5.3)

Review: The grams of a substance is determined using its molar mass as a factor.

$$\text{moles of substance} \times \frac{\text{grams}}{\text{1 mole of substance}} = \text{grams}$$

C. Find the number of grams in the following mole quantities:

1. 0.100 mole SO_2
2. 0.100 mole of H_2SO_4
3. 2.50 moles of NH_3
4. 1.25 moles of O_2
5. 0.500 mole of magnesium
6. 5.00 moles of H_2
7. 10.0 moles of PCl_3
8. 0.400 mole of S

Answers: (1) 6.41 g (2) 9.81 g (3) 42.5 g (4) 40.0 g
(5) 12.2 g (6) 10.0 g (7) 1380 g (8) 12.8 g

Review: When the grams of a substance is given, the molar mass is used to calculate the number of moles of substance present.

$$\text{grams of substance} \times \frac{1 \text{ mole}}{\text{grams}} = \text{moles}$$

Example: How many moles of NaOH are in 4.0 g NaOH?

$$4.0 \text{ g NaOH} \times \frac{1 \text{ mole NaOH}}{40.0 \text{ g NaOH}} = 0.10 \text{ mole NaOH}$$

D. Calculate the number of moles in the following:

1. 32.0 g of CH_4
4. 391 g of K
2. 8.00 g of C_3H_8
5. 25.0 g of Cl_2
3. 0.220 g of CO_2
6. 5.00 g of Al_2O_3
7. How many moles of iron are in an iron ingot with a mass of 150.0 g?
8. If brass is 40.0% copper, how many moles of copper are in a brass key with a mass of 75.0 g?
9. The methane burned in a gas heater has a formula of CH_4. If 725 g of methane are used in 1 month, how many moles of methane were burned?
10. There are 18 mg of iron in a vitamin tablet. If there are 100 tablets in a bottle, how many moles of iron are contained in the vitamins in the bottle?

Answers: (1) 2.00 moles (2) 0.182 mole (3) 0.00500 mole (4) 10.0 moles
(5) 0.352 mole (6) 0.0490 mole (7) 2.69 moles (8) 0.472 mole
(9) 45.3 moles (10) 0.032 mole

CHEMICAL CHANGE (5.4)

E. Identify each of the following as a chemical (C) or a physical (P) change:

1. _____ tearing a piece of paper
2. _____ burning paper
3. _____ rusting iron
4. _____ digestion of food
5. _____ dissolving salt in water
6. _____ boiling water
7. _____ chewing gum
8. _____ removing tarnish with silver polish

Answers 1. P 2. C 3. C 4. C 5. P 6. P 7. P 8. C

CHEMICAL EQUATIONS (5.5)

F. Write a balanced equation for the following statements of chemical reactions:

1. One unit of calcium carbonate when heated forms one unit calcium oxide and one molecule of carbon dioxide.

2. Two atoms of sodium metal react with one molecule of water to form one unit of sodium oxide and one molecule of hydrogen (H_2).

3. Two units of silver nitrate reacts with one unit potassium sulfide to form two units of potassium nitrate and one unit of silver sulfide.

4. Two units of aluminum hydroxide react with three molecules of sulfuric acid (H_2SO_4) to form one unit of aluminum sulfate and six molecules of water.

5. One molecule pentane (C_5H_{12}) reacts with eight molecules of oxygen (O_2) to produce five molecules of carbon dioxide and six molecules of water.

Answers:

(1) $CaCO_3 \longrightarrow CaO + CO_2$

(2) $2\ Na + H_2O \longrightarrow Na_2O + H_2$

(3) $2\ AgNO_3 + K_2S \longrightarrow 2\ KNO_3 + Ag_2S$

(4) $2\ Al(OH)_3 + 3\ H_2SO_4 \longrightarrow Al_2(SO_4)_3 + 6\ H_2O$

(5) $C_5H_{12} + 8\ O_2 \longrightarrow 5\ CO_2 + 6\ H_2O$

BALANCING CHEMICAL EQUATIONS (5.6)

G. Place the appropriate coefficient (including 1) in front of each formula to show a balanced equation.

1. ___MgO ⟶ ___Mg + ___O_2
2. ___ Zn + ___ HCl ⟶ ___ $ZnCl_2$ + ___ H_2
3. ___ Al + ___ $CuSO_4$ ⟶ ___ Cu + ___ $Al_2(SO_4)_3$
4. ___ Al_2S_3 + ___ H_2O ⟶ ___ $Al(OH)_3$ + ___ H_2S
5. ___ $BaCl_2$ + ___ Na_2SO_4 ⟶ ___ $BaSO_4$ + ___ $NaCl$
6. ___ CO + ___ Fe_2O_3 ⟶ ___ Fe + ___ CO_2
7. ___ K + ___ H_2O ⟶ ___ K_2O + ___ H_2
8. ___ $Fe(OH)_3$ ⟶ ___ Fe_2O_3 + ___ H_2O

Answers:
(1) $2\ MgO \longrightarrow 2\ Mg + 1\ O_2$
(2) $1\ Zn + 2\ HCl \longrightarrow 1\ ZnCl_2 + 1\ H_2$
(3) $2\ Al + 3\ CuSO_4 \longrightarrow 3\ Cu + 1\ Al_2(SO_4)_3$
(4) $1\ Al_2S_3 + 6\ H_2O \longrightarrow 2\ Al(OH)_3 + 3\ H_2S$
(5) $1\ BaCl_2 + 1\ Na_2SO_4 \longrightarrow 1\ BaSO_4 + 2\ NaCl$
(6) $3\ CO + 1\ Fe_2O_3 \longrightarrow 2\ Fe + 3\ CO_2$
(7) $2\ K + 1\ H_2O \longrightarrow 1\ K_2O + 1\ H_2$
(8) $2\ Fe(OH)_3 \longrightarrow 1\ Fe_2O_3 + 3\ H_2O$

MOLE RELATIONSHIPS IN CHEMICAL EQUATIONS (5.7)

Review: The coefficients in a balanced chemical equation can be used to express the mole relationship between any two substances in that equation. For the equation, A ⟶ B, we can write the following factors:

$$\frac{\text{(coefficient) moles A}}{\text{(coefficient) moles B}} \quad \text{and} \quad \frac{\text{(coefficient) moles B}}{\text{(coefficient) moles A}}$$

H. Write the conversion factors that are possible from the following equations:

$N_2 + O_2 \longrightarrow 2\ NO$

Answers: 1.

$$\frac{1\text{ mole }N_2}{1\text{ mole }O_2} \text{ and } \frac{1\text{ mole }O_2}{1\text{ mole }N_2}; \quad \frac{1\text{ mole }N_2}{2\text{ mole NO}} \text{ and } \frac{2\text{ mole NO}}{1\text{ mole }N_2}$$

$$\frac{1\text{ mole }O_2}{2\text{ mole NO}} \text{ and } \frac{2\text{ mole NO}}{1\text{ mole }O_2}$$

CALCULATIONS USING EQUATIONS (5.8)

Review: Use the appropriate mole factors to change from the number of moles of the given substance to moles of the desired products. Using the equation N_2 + P_2 ⟶ 2NO, calculate the moles of NO obtained from 3 moles N_2. The solution uses the following set up with a mole factor using the coefficients of the substances in the equation.

$$3 \text{ moles } N_2 \times \underbrace{\frac{2 \text{ mole NO}}{1 \text{ mole } N_2}}_{\text{mole factor}} = 6 \text{ moles NO}$$

I. Consider the equation $2\,Na + H_2O \longrightarrow Na_2O + H_2$.

1. How many moles of water are needed to react with 4.00 moles of sodium?

2. How many moles of sodium will react with 18.0 g of H_2O?

3. How many grams of sodium are needed to produce 2.00 moles Na_2O?

4. How many grams of water are need to produce 31.0 g Na_2O?

5. How many grams of hydrogen are produced with 36.0 g of H_2O react?

6. When 3.0 moles of Na reacts, how many grams of H_2 are produced?

Answers: (1) 2.00 moles H_2O (2) 2.00 moles Na (3) 92.0 g Na

(4) 9.00 g H_2O (5) 4.00 g H_2 (6) 3.0 g H_2

SELF-EVALUATION TEST

CHEMICAL QUANTITIES AND REACTIONS

1. What is the molar mass of Li_2SO_4?
A. 55.1 g B. 62.1 g C. 100.1 g D. 109.9 g E. 103.1 g

2. What is the molar mass of $NaNO_3$?
A. 34.0 g B. 37.0 g C. 53.0 g D. 75.0 g E. 85.0 g

3. The number of grams in 0.600 mole of Cl_2 is
A. 71.0 g B. 118 g C. 42.6 g D. 84.5 g E. 4.30 g

4. How many grams are in 4.00 moles of NH_3?
A. 4.00 g B. 17.0 g C. 34.0 g D. 68.0 g E. 0.240 g

5. How many grams are in 4.50 moles of N_2?
A. 14.0 g. B. 28.0 g C. 56.0 g D. 112 g E. 126 g

6. How many moles is 8.0 g NaOH?
A. 0.10 mole B. 0.20 mole C. 0.40 mole D. 2.0 moles E. 4.0 moles

7. The number of moles of aluminum in 54 g of Al is
A. 0.50 mole B. 1.0 mole C. 2.0 moles D. 3.0 moles E. 4.0 moles

8. The number of moles of water in 36 g of H_2O is
A. 0.50 mole B. 1.0 mole C. 2.0 moles D. 3.0 moles E. 4.0 moles

9. What is the number of moles in 2.2 g CO_2?
A. 2.0 moles B. 1.0 moles C. 0.20 mole D. 0.050 mole E. 0.010 mole

10. 0.20 g H_2 = ____mole H_2
A. 0.10 mole B. 0.20 mole C. 0.050 mole D. 0.020 mole E. 0.010 mole

Indicate whether each change is a A. physical change or a B. chemical change:

11. a melting ice cube

12. breaking glass

13. bleaching a stain

14. a burning candle

15. milk turning sour

For questions 16, 17, 18, 19, and 20, balance the equations and indicate the correct coefficient for the underlined component.

A. 1 B. 2 C. 3 D. 4 E. 5

16. $Sn + \underline{Cl_2} \longrightarrow SnCl_4$

17. $Al + H_2O \longrightarrow Al_2O_3 + \underline{H_2}$

18. $C_3H_8 + \underline{O_2} \longrightarrow CO_2 + H_2O$

19. $\underline{NH_3} + O_2 \longrightarrow N_2 + H_2O$

20. $N_2O \longrightarrow N_2 + \underline{O_2}$

For questions, 21, 22, 23. 24. amd 25, consider the reaction

$$\underset{\text{ethanol}}{C_2H_5OH} + 3O_2 \longrightarrow 2CO_2 + 3H_2O$$

21. How many grams of oxygen are needed to react with 1.0 mole of ethanol?
A. 8.0 g B. 16 g C. 32 g D. 64 g E. 96 g

22. How many moles of water are produced when 12 moles of oxygen react?
A. 3.0 moles B. 6.0 moles C. 8.0 moles D. 12 moles E. 36 moles

23. How many grams of carbon dioxide are produced when 92 g of ethanol react?
A. 22 g B. 44 g C. 88 g D. 92 g E. 176 g

24. How many moles of oxygen would be needed to produce 44 g of CO_2?
A. 0.67 moles B. 1.0 mole C. 1.5 moles D. 2.0 moles E. 3.0 moles

25. How many grams of water will be produced if 23 g of ethanol react?
A. 54 g B. 27 g C. 18 g D. 9.0 g E. 6.0 g

ANSWERS FOR THE SELF-EVALUATION TEST

1. D	6. B	11. A	16. B	21. E
2. E	7. C	12. A	17. C	22. D
3. C	8. C	13. B	18. E	23. E
4. D	9. D	14. B	19. D	24. C
5. E	10.A	15. B	20. A	25. B

SCORING THE SELF-EVALUATION TEST

25 questions 4 points each 100 points total

CHAPTER 6
NUCLEAR RADIATION

KEY CONCEPTS

1. Radiation emitted from the nucleus of an unstable atom may be in the form of alpha particles, beta particles, or gamma rays.

2. When radiation is emitted from a nucleus, there is a change in the particles or energy of the nucleus.

3. Nuclear radiation is capable of severely damaging or destroying cells in the body; therefore, protection such as shielding, distance, and time must be used wherever possible to avoid undesired exposure to radiation.

4. Radioactive decay is described in terms of a half-life which is the amount of time required for one-half of the sample to decay.

5. Nuclear fission and nuclear fusion are two ways of releasing large amounts of energy from the nuclei of atoms.

KEY WORDS

Using complete sentences, write a description for each of the following key words:

radioactive isotope

alpha particle

beta particle

gamma ray

LEARNING EXERCISES

Radioactivity (6.1)

A. Write the nuclear symbol for the following types of radiation:

1. alpha particle ________
2. beta particle ________
3. gamma ray ________

Answers: 1. $^{4}_{2}He$, α 2. $^{0}_{-1}e$, ß 3. γ

B. Match the description in column B with the terms in column A:

A		B
1. ____ $^{18}_{8}O$	a.	Symbol for a beta particle
2. ____scan	b.	Time required for one-half of a radioactive sample to decay
3. ____rad	c.	Symbol for an alpha particle
4. ____fission	d.	Process whereby radiation is emitted by an unstable nucleus
5. ____$^{4}_{2}He$	e.	Nuclear symbol for an atom of oxygen
6. ____radioisotope	f.	An image of an organ made through detection of radioactivity
7. ____ß	g.	Symbol for gamma radiation
8. ____half-life	h.	Breakdown of a large nucleus with the release of energy
9. ____ γ	i.	Radioactive form of an element
10.____decay	j.	A unit of radiation measurement

682-8540

3772 Tibbets Riverside CA

Answers: 1. e 2. f 3. j 4. h 5. c 6. i 7. a 8. b 9. g 10. d

C. *ESSAY:* Discuss three ways in which you can minimize the amount of radiation received when working with radioactive materials.

Answer: Three ways to minimize exposure to radiation are to use shielding, to shorten time in the radioactive area, and to keep as much distance as possible from the radioactive materials.

D. State the type(s) of radiation that could use each of the following shielding materials for protection.

1. clothing ____________________
2. your skin ____________________
3. paper ____________________
4. thick concrete ____________________
5. lead wall ____________________

Answers: 1. alpha, beta 2. alpha 3. alpha, beta 4. alpha, beta, gamma 5. alpha, beta, gamma

NUCLEAR EQUATIONS (6.2)

E. Write a nuclear symbol that completes each of the following nuclear equations:

1. $^{66}_{29}Cu \longrightarrow {}^{66}_{30}Zn + ?$

2. $^{127}_{53}I \longrightarrow {}^{1}_{0}n + ?$

3. $^{238}_{92}U \longrightarrow {}^{4}_{2}He + ?$

4. $^{24}_{11}Na \longrightarrow {}^{0}_{-1}e + ?$

5. $? \longrightarrow {}^{30}_{14}Si + {}^{0}_{-1}e$

Answers 1. $^{0}_{-1}e$ 2. $^{126}_{53}I$ 3. $^{234}_{90}Th$ 4. $^{24}_{12}Mg$ 5. $^{30}_{13}Al$

DETECTING AND MEASURING RADIATION (6.3)

F. Define the following units used to measure radiation.

Curie

rad

rem

Answers: A **Curie** is a quantity of radioactive isotope that gives 3.7×10^{10} disintegrations per second.
A **rad** (10^{-5} J/g) measures the absorption of radiation by body tissue.
A **rem** measures the biological damage (rad x RBE) caused by the absorption of radiation by the tissues.

MEDICAL APPLICATIONS OF RADIOISOTOPES (6.4)

G. Write the nuclear symbol for each of the following radioisotopes:

1. The radioisotope iodine-131 is an ideal tracer for studying thyroid gland activity. ________

2. Phosphorus-32 has been used to locate brain tumors because the cells in these tumors grow rapidly. ________

3. Sodium-24 may be used to determine blood flow. It can be used to locate a blood clot or embolism. ________

4. Nitrogen-13 is used in positron emission tomagraphy. ________

Answers 1. ${}^{131}_{53}I$ 2. ${}^{32}_{15}P$ 3. ${}^{24}_{11}Na$ 4. ${}^{13}_{7}N$

HALF-LIFE OF A RADIOISOTOPE (6.5)

H. Calculate the quantities of radioisotope that are active in the following:

1. ^{125}I has a half-life of 60 days. If you have 80 mg of I-125, how many mg will still be radioactive

   a. after one half-life?

   b. after two half-lives?

   c. after 240 days?

2. ^{99m}Tc has a half-life of 6 hours. If a radiation technician picked up a 16 mg sample at 8 AM, how much of the radioactive sample remained at 8 PM that same day?

3. Phosphorus-32 has a half-life of 14 days. How much of a 200 μg sample will be radioactive after 56 days?

4. Iodine-131 has a half life of 8 days. How many days are needed for 80 mg sample to decay to 5 mg?

Answers: (1.) a. 40 mg b. 20 mg c. 5 mg (2.) 4 mg (3.) 12.5 μg (4.) 32 days

PRODUCING RADIOISOTOPES (6.6)

F. Complete the following equations for bombardment reactions:

1. $^{40}_{20}Ca$ + ? ⟶ $^{40}_{19}K$ + $^{1}_{1}H$

2. $^{27}_{13}Al$ + $^{1}_{0}n$ ⟶ $^{24}_{11}Na$ + ?

3. $^{10}_{5}B$ + $^{1}_{0}n$ ⟶ $^{4}_{2}He$ + ?

4. $^{23}_{11}Na$ + ? ⟶ $^{23}_{12}Mg$ + $^{1}_{0}n$

5. $^{197}_{79}Au$ + $^{1}_{1}H$ ⟶ ? + $^{1}_{0}n$

Answers 1. $^{1}_{0}n$ 2. $^{4}_{2}\alpha$ 3. $^{7}_{3}Li$ 4. $^{1}_{1}H$ 5. $^{197}_{80}Hg$

NUCLEAR FISSION AND FUSION (6.7)

I. *ESSAY*: Compare the method of producing energy by fission and fusion.

Essay Answer: Nuclear fission is a splitting of the atom into two or more nuclei accompanied by the release of large amounts of energy and radiation. Nuclear fusion is the combining of two or more nuclei to form a heavier nucleus which causes a release of a large amount of energy. However, fusion requires a considerable amount of energy to initiate the reaction.

SELF-EVALUATION TEST
NUCLEAR RADIATION

1. The correctly written nuclear symbol is

   A. $^{30}_{14}Si$ B. $^{14}_{30}Si$ C. $^{30}_{14}Si$ D. $^{30}_{16}Si$ E. $^{16}_{30}Si$

2. Alpha particles are composed of
   A. protons B. neutrons C. electrons
   D. protons and electrons E. protons and neutrons

3. Beta particles are composed of
   A. protons B. neutrons C. electrons
   D. protons and electrons E. protons and neutrons

4. To provide shielding from gamma rays, which is needed?
   A. skin B. paper C. heavy clothing D. lead E. air

5. The skin will provide shielding from
   A. alpha particles B. beta particles C. gamma rays
   D. ultraviolet rays E. x rays

6. When beta particles are emitted, atoms become
   A. heavier B. lighter C. brighter D. more stable E. less stable

7. Gamma radiation is a type of radiation that
   A. originates in the electron shells.
   B. is most dangerous.
   C. is least dangerous.
   D. is the heaviest.
   E. goes the shortest distance.

8. The unit used to measure the number of disintegrations per second in a radioactive sample is called a
   A. Curie B. rad C. rem D. RBE E. MDP

9. The radioisotope iodine-131 is used as a radioactive tracer for studying thyroid gland activity. The nuclear symbol for iodine-131 is

   A. I B. $_{131}I$ C. $^{131}_{53}I$ D. ^{53}I E. $^{78}_{53}I$

10. What completes the following reaction?

$$^{14}_{7}N + {}^{1}_{0}n \longrightarrow ? + {}^{1}_{1}H$$

   A. $^{15}_{8}O$ B. $^{15}_{6}C$ C. $^{14}_{8}O$ D. $^{14}_{6}C$ E. $^{15}_{7}N$

11. An atom undergoing decay by emitting an alpha particle will
    A. increase its mass by 1.
    B. increase its mass by 2.
    C. increase its mass by 4.
    D. decrease its mass by 2.
    E. decrease its mass by 4.

12. To complete this nuclear equation, you need to write

$$^{54}_{26}Fe + ? \longrightarrow {}^{57}_{28}Ni + {}^{1}_{0}n$$

    A. an alpha particle
    B. a beta particle
    C. gamma
    D. neutron
    E. proton

13. The time required for a radioisotope to decay is measured by its
    A. half-life
    B. protons
    C. nuclear symbol
    D. radioisotope
    E. fusion

14. Oxygen-15 used in PET imaging has a half-life of 2 min. How many half-lives have occurred in the 10 minutes it takes to prepare the sample?
    A. 2 B. 3 C. 4 D. 5 E. 6

15. Iodine-131 has a half-life of 8 days. How many days will it take for a 160 mg sample to decay to 10 mg?
    A. 8 days B. 16 days C. 32 days D. 40 days E. 48 days

16. Phosphorus-32 has a half-life of 14 days. After 28 days, how many mg of a 100 mg sample will still be radioactive?
    A. 75 mg B. 50 mg C. 40 mg D. 25 mg E. 12.5 mg

17. An imaging technique that uses the energy emitted by exciting the nuclei of hydrogen atoms is called
    A. computerized tomography (CT)
    B. positron emission tomography (PET)
    C. radioactive tracer
    D. magnetic resonance imaging (MRI)
    E. radiation

18. An imaging technique that detects the absorption of x-rays by the body tissues is called

A. computerized tomography (CT)
B. positron emission tomography (PET)
C. radioactive tracer
D. magnetic resonance imaging (MRI)
E. radiation

19. The "splitting" of a large nucleus to form smaller particles accompanied by a release of energy is called

A. radioisotope B. fission C. fusion D. rem E. half-life

20. The fusion reaction

A. occurs in the sun.
B. forms larger nuclei from smaller nuclei.
C. requires extremely high temperatures.
D. releases a large amount of energy.
E. All of the above

ANSWERS TO THE SELF-EVALUATION TEST

1. A	6. D	11. E	16. D
2. E	7. B	12. A	17. D
3. C	8. A	13. A	18. A
4. D	9. C	14. D	19. B
5. A	10.D	15. C	20. E

SCORING THE SELF-EVALUATION TEST

20 questions 5 points each 100 points total

CHAPTER 7
GASES

KEY CONCEPTS

1. The kinetic molecular theory describes gases as composed of small particles with great distances between them, moving at high speeds until they collide with each other or the walls of the container.

2. A description of a gas includes the temperature, volume, pressure, and number of moles of the gas.

3. Boyle's law states that the pressure and volume of a gas are inversely related when temperature is constant.

4. Charles' law states that the volume of a gas is directly related to the absolute temperature when pressure remains constant.

5. Gay-Lussac's law states that the pressure of a gas is directly related to the absolute temperature when volume is constant.

6. Using the combined gas law, the pressure, volume, or temperature of a gas can be found from values given for the other two variables.

7. One mole of any gas occupies a volume of 22.4 L at STP (0°C and 1 atm).

8. According to Dalton's law, the partial pressure of a gas in a gas mixture is the pressure that gas exerts if it were the only gas in the container.

9. The ideal gas law is a relationship between the pressure, volume, moles and absolute temperature of gas related by the universal gas constant.

KEY WORDS

Use complete sentences to write a description for each of the following:

kinetic molecular theory

pressure

Boyle's law

Charles' law

partial pressure

LEARNING EXERCISES

PROPERTIES OF GASES (7.1)

A. True or false

1. ____Gases are composed of small particles.
2. ____Gas molecules are usually close together.
3. ____Gas molecules move rapidly because they are strongly attracted.
4. ____The distances between gas molecules are great.
5. ____Gas molecules travel in straight lines until they collide.

Answers: 1. T 2. F 3. F 4. T 5. T

GAS PRESSURE

B. Complete the following:

1. 1.5 atm = ________ torr
2. 550 mm Hg = ________ atm
3. 725 mm Hg = ________ atm
4. 1520 torr = ________ atm
5. 30.5 inches Hg = ________ torr

Answers: (1) 1140 torr (2) 0.724 atm (3) 725 torr (4) 2.00 atm (5) 775 mm Hg

C. ESSAY: Explain how a liquid can have a boiling point of 80°C at sea level and a boiling point of 74°C at an altitude of 1000 m?

Essay Answer: The vapor pressure required for boiling is lower at higher altitude because the atmospheric pressure is lower. As a result, the substance boils at a lower temperature.

PRESSURE AND VOLUME - BOYLE'S LAW (7.3)

Review: Boyle's Law states that the volume of a gas changes inversely with the pressure when temperature remains constant.

D. Complete with ***increases*** or ***decreases***:

1. Gas pressure increases (T constant) when volume ____________.
2. Gas volume increases at constant T when pressure ____________

Answers: 1. decreases 2. decreases

E. Calculate the variable in each of the following gas problems using ***Boyle's Law***.

1. Four (4.0) liters of helium gas have a pressure of 800.0 torr. What will be the new pressure if the volume is reduced to 1.0 liter (n and T constant)?

2. A gas occupies a volume of 360 mL at 750 torr. What volume will it occupy when the pressure is changed to (a) 1500 torr? (b) 375 torr? (n and T constant)

3. A gas sample under a pressure of 5.0 atm has volume of 3.00 liters. If the gas pressure is changed to 760 torr, what volume will the gas occupy (n and T constant)?

Answers: (1) 3200 torr (2) a. 180 mL b. 720 mL (3) 15 L

TEMPERATURE AND VOLUME - CHARLES' LAW (7.4)

Review: Charles' Law states that the volume of gas varies directly with the Kelvin temperature when pressure remains constant.

F. Complete with ***increases*** or ***decreases***:

1. When temperature increases at constant pressure, volume ___________.

2. When volume decreases at constant pressure, temperature ___________.

Answers: 1. increases 2. decreases

G. Use ***Charles' Law*** to solve the following gas problems:

1. A large balloon has a volume of 2.5 L at a temperature of 0°C. What is the new volume of the balloon when the temperature rises to 120°C when P remains constant?

2. Consider a balloon filled with helium to a volume of 6600 L at a temperature of 223°C. To what temperature must the gas be cooled to decrease the volume to 4800 L (P constant)?

Answers: (1) 3.6 L (2) 88°C

TEMPERATURE AND PRESSURE - GAY-LUSSAC'S LAW (7.5)

Review: Gay-Lussac's Law states that the pressure of a gas varies directly with the Kelvin temperature when volume remains constant.

H. Solve the following gas law problems using ***Gay-Lussac's Law***.

1. A sample of helium gas has a pressure of 860 mm Hg at a temperature of 225 K. What pressure (mm Hg) will the helium have when the temperature of the sample reaches 675 K (V constant)?

2. A balloon contains a gas with a pressure of 580 mm Hg and a temperature of 227°C. What is the new pressure (mm Hg) of the gas when the temperature drops to 27°C (V constant)?

3. A spray can contains a gas with a pressure of 3.0 atm at a temperature 17°C. What is the pressure (atm) inside the container if the temperature inside the can rises to 110°C (V constant)?

Answers: (1) 2580 mm Hg (2) 348 mm Hg (3) 4.0 atm

THE COMBINED GAS LAW OF PRESSURE, VOLUME, AND TEMPERATURE RELATIONSHIPS (7.6)

I. Fill in the blanks by writing ***increases*** or ***decreases*** for a gas in a closed container.

	Pressure	***Volume***	***Moles***	***Temperature***
1.	________	increases	constant	constant
2.	increases	constant	increases	constant
3.	constant	decreases	decreases	constant
4.	in-	constant	constant	increases
5.	constant	________	constant	decreases
6.	________	constant	increases	constant
7.	decreases	________	constant	constant
8.	decreases	constant	________	constant

Answers: 1. decreases 2. increases 3. decreases 4. increases
5. decreases 6. increases 7. increases 8. decreases

J. Solve the following using the combined gas laws:

1. A 5.0 L sample of nitrogen gas has a pressure of 1200 torr. What is the new pressure if the temperature is changed from 220K to 440 K and the volume increased to 20.0 liters?

2. A 10.0 L sample of gas is emitted from a volcano with a pressure of 1.20 atm and a temperature of 150°C. What is the volume of the gas when the pressure near the volcano is 0.90 atm and the temperature is −40°C?

3. A 25 mL bubble forms at the ocean depths where the pressure is 10.0 atm and the temperature is 5.0°C. What is the volume of that bubble at the ocean surface where the pressure is 760 mm Hg and the temperature is 25°C?

Answers: (1) 600 torr (2) 7.3 L (3) 268 mL

VOLUME AND MOLES - AVOGADRO'S LAW (7.7)

***Review*:** Avogadro's Law states that equal volumes of gases at the same temperature and pressure contain the same number of molecules.

K. Use Avogadro's Law to solve the following gas problems:

1. A gas containing 0.50 mole of helium has a volume of 4.00 L. What is the new volume when 1.0 mole of nitrogen is added to the container when pressure and temperature remain constant?

2. A balloon containing 1.00 mole of oxygen has a volume of 15 L. What is the new volume of the balloon when 2.00 moles of helium are added (T and P constant)?

3. What is the volume occupied by 28.0 g of nitrogen (N_2) at STP?

4. What is the volume (L) of a container that holds 6.40 g O_2 at STP?

Answers: (1) 12.0 L (2) 45.0 L (3) 44.8 L (4) 4.48 L

PARTIAL PRESSURE - DALTON'S LAW (7.8)

L. Use Dalton's law to solve the following problems about gas mixtures:

1. What is the pressure in mm Hg of a sample of gases containing oxygen at 0.500 atm, nitrogen (N_2) at 132 torr, and helium at 224 mm Hg?

2. What is the pressure (atm) of a gas sample containing helium at 285 torr and oxygen (O_2) at 1.20 atm?

3. A gas sample containing nitrogen (N_2) and oxygen (O_2) has a pressure of 1500 mm Hg. If the partial pressure of the nitrogen is 0.900 atm, what is the partial pressure (mm Hg) of the oxygen gas in the mixture?

Answers: (1) 736 mm Hg (2) 1.58 atm (3) 816 mm Hg

M. Complete the table for typical blood gas values for partial pressures.

Gas	*alveoli*	*oxygenated blood*	*deoxygenated blood*	*tissues*
CO_2	________	________	________	________
O_2	________	________	________	________

Answers: CO_2: 40 torr; 40 torr; 46 torr; 50 torr
O_2: 100 torr; 100 torr; 40 torr; 30 torr

THE IDEAL GAS LAW (7.9)

Review: The ideal gas law gives the relationship between the four gas variables: PV = nRT
R is the universal gas constant = 0.0821 L atm/mole °K

N. Use the ideal gas law to solve for the unknown variable in each of the following:

1. What volume (L) will be occupied by 0.25 mole of nitrogen gas at 0°C and 1.50 atm?

2. What is the temperature (°C) of 0.500 mole helium that occupies a volume of 15.0 L at a pressure of 1200 mm Hg?

3. What is the pressure in atm of 1.0 mole of neon in a 5.0 L steel container at a temperature of 18°C?

Answers: (1) 3.7 L (2) 304°C (3) 4.8 atm

SELF-EVALUATION TEST

GASES

Answer questions 1-5 using true (A) or false (B):

1. ____ A gas does not have its own volume or shape.

2. ____ The molecules of a gas are moving extremely fast.

3. ____ The collisions of gas molecules with the walls of their container create pressure.

4. ____ Gas molecules are close together and move in straight-line patterns.

5. ____ We consider gas molecules to have no attractions between them.

6. When a gas is heated in a closed metal container, the
   A. pressure increases.
   B. pressure decreases.
   C. volume increases.
   D. volume decreases.
   E. number of molecules increases.

7. The pressure of a gas will increase when
   A. the volume increases.
   B. the temperature decreases.
   C. more molecules of gas are added.
   D. molecules of gas are removed.
   E. None of these

8. If the temperature of a gas is increased,
   A. the pressure will decrease.
   B. the volume will increase.
   C. the volume will decrease.
   D. the number of molecules will increase.
   E. None of these

9. The relationship that the volume of a gas is inversely related to its pressure at constant temperature is known as
   A. Boyle's Law
   B. Charles' Law
   C. Gay-Lussac's Law
   D. Dalton's Law
   E. Avogadro's Law

10. What is the pressure in atmospheres of a gas pressure of 1200 mm Hg?
   A. 0.63 atm B. 0.79 atm C. 1.2 atm D. 1.6 atm E. 2.0 atm

11. A 6.00 L sample of oxygen has a pressure of 660.0 mm Hg. When the volume is reduced to 2.00 liters at constant temperature, it will have a new pressure of

   A. 1980 mm Hg B. 1320 mm Hg C. 330.0 mm Hg
   D. 220.0 mm Hg E. 110.0 mm Hg

12. A sample of nitrogen gas at 180 K has a pressure of 1.0 atm. When the temperature is increased to 360 K at constant volume, the new pressure will be

   A. 0.50 atm B. 1.0 atm C. 1.5 atm D. 2.0 atm E. 4.0 atm

13. If two gases have the same volume, temperature, and pressure, they also have the same
   A. density
   B. number of molecules
   C. molar mass
   D. speed
   E. size molecules

14. A gas sample with a volume of 4.00 L has a pressure of 750 mm Hg and a temperature of 77°C. What is its new volume at 277°C and 250 mm Hg?

A. 7.6 L B. 18.9 L C. 2.1 L D. 0.00056 L E. 3.3 L

15. If the temperature of a gas does not change, but its volume doubles, its pressure will
A. double.
B. triple.
C. decrease to one-half the original pressure.
D. decrease to one-fourth the original pressure.
E. not change.

16. A sample of oxygen with a pressure of 400 mm Hg contains 2.0 moles of gas and has a volume of 4.0 L. What will the new pressure be when the volume expands to 5.0 L, and 3.0 moles of helium gas are added while temperature is constant?
A. 160 mm Hg B. 250 mm Hg C. 800 mm Hg
D. 1000 mm Hg E. 1560 mm Hg

17. What is the pressure in atm of a gas that has a volume of 2.50 L, a temperature of 27°C and contains 2.00 moles of the gas?
A. 1.77 atm
B. 0.0508 atm
C. 2.21 atm
D. 19.7 atm
E. 0.00137 atm

18. The conditions for standard temperature and pressure (STP) are
A. 0 K, 1 atm
B. 0°C, 10 atm
C. 25°C, 1 atm
D. 273 K, 1 atm
E. 273 K, 0.5 atm

19. The volume occupied by 1.50 moles of CH_4 at STP is

A. 44.8 L B. 33.6 L C. 22.4 L D. 11.2 L E. 5.60 L

20. How many grams of oxygen gas (O_2) are present in 44.8 L of oxygen at STP?

A. 8.0 g B. 16.0 g C. 32.0 g D. 48.0 g E. 64.0 g

21. A gas mixture contains helium with a partial pressure of 0.80 atm, oxygen with a partial pressure of 450 mm Hg, and nitrogen with a partial pressure of 230 torr. What is the total pressure in atm for the gas mixture?

A. 1.10 atm B. 1.39 atm C. 1.69 atm D. 2.00 atm E. 8.00 atm

22. A mixture of oxygen and nitrogen has a total pressure of 840 mm Hg. If the oxygen has a partial pressure of 510 mm Hg, what is the partial pressure of the nitrogen?
A. 240 mm Hg
B. 330 mm Hg
C. 775 mm Hg
D. 1040 mm Hg
E. 1350 mm Hg

23. The exchange of gases between the alveoli, blood, and tissues of the body is a result of
A. pressure gradients.
B. different molecular weights.
C. shapes of molecules.
D. altitude.
E. All of these

24. Oxygen moves into the tissues from the blood because its partial pressure
A. in arterial blood is higher than in the tissues.
B. in venous blood is higher than in the tissues.
C. in arterial blood is lower than in the tissues.
D. in venous blood is lower than in the tissues.
E. is equal in the blood and in the tissues.

25. What is the volume in liters of 0.50 moles of $N_2(g)$ at a temperature of 25°C and a pressure of 2.0 atm?
A. 0.51 L
B. 1.0 L
C. 4.2 L
D. 6.1 L
E. 24 L

ANSWERS TO THE SELF-EVALUATION TEST

1. A	6. A	11. A	16. C	21. C
2. A	7. C	12. D	17. D	22. B
3. A	8. B	13. B	18. D	23. A
4. B	9. A	14. B	19. B	24. A
5. A	10.D	15. C	20. E	25. D

SCORING THE SELF-EVALUATION TEST

25 questions 4 points each 100 points total

CHAPTER 8
SOLUTIONS

KEY CONCEPTS

1. Solutions are composed of solutes and solvents.

2. The polarity of the water molecules allows hydrogen bonding.

3. In general, solutes dissolve in solvents that are similar in polarity. When ionic substances dissolve in water, they separate into positive and negative ions.

4. A solution is saturated when it contains the maximum amount of dissolved solute (solubility) at a given temperature. The rate of dissolution is affected by the size of the solute particles, the temperature of the solvent, and by agitation.

5. The concentration of a solution states a relationship of the quantity of solute in a given volume of the solution. The percent concentration indicates the grams of solute per 100 mL of solution.

6. The molarity of a solution indicates the moles of solute in 1 L of solution.

7. When a solution is diluted with water, the concentration decreases.

8. The size of the solute particles will determine whether a mixture is a solution, a colloid, or a suspension. A semipermeable membrane retains particles that are colloids or larger.

9. Osmosis involves the movement of water towards an area of higher solute concentration. Osmotic pressure increases as the number of solute particles increases. An isotonic solution has the same osmotic pressure as the body fluids.

KEY WORDS

Using complete sentences, describe each of the following key terms:

solution

percent concentration

molarity

osmosis

LEARNING EXERCISES

SOLUTES AND SOLVENTS (8.1)

A. Indicate the solute and solvent in each of the following solutions:

Solute	*Solvent*	
________	________	1. 10 g KCl dissolved in 100 g of water.
________	________	2. Soda water: $CO_2(g)$ dissolved in water.
________	________	3. An alloy composed of 80% Zn and 20% Cu.
________	________	4. A mixture of O_2 (200 mm Hg) and He (500 mm Hg).
________	________	5. A solution of 40 mL CCl_4 and 2 mL Br_2.

Answers: 1. KCl; water 2. CO_2; water 3. Cu; Zn 4. oxygen; helium 5. Br_2; CCl_4

WATER: AN IMPORTANT SOLVENT (8.2)

B. *ESSAY*: How does the polarity of the water molecule allow it to hydrogen bond?

Essay Answer: The hydrogen-to-oxygen bonds in water molecules are polar because the hydrogen atoms are partially positive and the oxygen atoms are partially negative. Hydrogen bonding occurs because the positive hydrogen atoms in one water molecules are attracted to negative oxygen atoms of other water molecules.

FORMATION OF SOLUTIONS (8.3)

C. Water is polar and hexane is nonpolar. In which of these two solvents would the following substances be *more* soluble?

___________ 1. bromine, Br_2, nonpolar

___________ 2. HCl, polar

___________ 3. cholesterol, nonpolar

___________ 4. vitamin D, nonpolar

___________ 5. vitamin C, polar

Answers: 1. hexane 2. water 3. hexane 4. hexane 5. water

SATURATIONS (8.4)

D. Indicate whether the following statements describe a saturated solution (S) or an unsaturated solution (U):

_____1. A sugar cube dissolves when added to a cup of coffee.

_____2. A KCl crystal added to a KCl solution does not change in size.

_____3. A layer of sugar forms in the bottom of a glass of ice tea.

_____4. The rate of crystal formation equals the rate of solution.

_____5. Upon heating, all the sugar dissolves.

Answers 1. U 2. S 3. S 4. S 5. U

E. Use the KNO_3 solubility chart for the following problems:

Solubility of KNO_3

Temperature (°C)	*$gKNO_3/100\ gH_2O$*
0	15
20	30
40	65
60	110
80	170
100	250

1. How many grams of KNO_3 will dissolve in 100 g of water at 40°C?

2. How many grams of KNO_3 will dissolve in 300 g of water at 60°C.

3. A solution is prepared using 200 g of water and 350 g of KNO_3 at 80°C. Will any solute remain undissolved? If so, how much?

4. Will 200 g of solute dissolve when added to 100 g of water at 100°C?

Answers: (1) 65 g (2) 330 g (3) Yes. 10 g of KNO_3 will not dissolve.
(4) Yes, all 200 g of solute will dissolve.

F. Indicate whether each of the following solution preparations increases or decreases the rate of solution:

1.__________ Chilling the solvent before adding solute.

2.__________ Crushing the solute.

3.__________ Heating the mixture.

4.__________ Using large chunks of solute.

5.__________ Stirring the mixture.

Answers: 1. decreases 2. increases 3. increases 4. decreases 5. increases

PERCENT CONCENTRATION (8.5)

Review: The concentration of a solution is the relationship (ratio) between the amount of solute to the total volume of solution which contains the solute. The quantity of solute is usually expressed in grams or moles.

$$\text{Percent concentration} = \frac{\text{grams of solute}}{\text{milliliters of solution}} \times 100$$

Example: What is the weight-volume % of 2.4 g of $NaHCO_3$ in 120 mL of $NaHCO_3$ solution?

Solution: $$\frac{2.4 \text{ g } NaHCO_3}{120 \text{ mL solution}} \times 100 = 2.0\ \%$$

G. Determine the percent concentration of the following solutions:

1. The weight/weight % of 18.0 g NaCl in 90.0 g of solution.

2. The weight/volume % of 5.0 g KCl in 2.0 liters of solution.

3. The weight/weight % of 1.0 g KOH in 25 g of solution.

4. The weight/volume % of 0.25 kg glucose in 5.0 liters of solution.

Answers: (1) 20.0% (2) 0.25% (3) 4.0% (4) 5.0%

Review: Use a percent concentration in factor form to solve concentration problems.

Weight-Weight Percent = $\dfrac{\text{Number of grams of solute}}{\text{in 100 g solution}}$

Weight-Volume Percent = $\dfrac{\text{Number of grams of solute}}{\text{in 100 mL of solution}}$

Volume-Volume Percent = $\dfrac{\text{Number of mL of solute}}{\text{in 100 mL of solution}}$

Example: Calculate the number of grams of KCl in 250 mL of 1.0% (w/v) KCl solution.

$$250 \text{ mL} \times \frac{1.0 \text{ g KCl}}{100 \text{ mL solution}} = 2.5 \text{ g KCl}$$

H. Calculate the number of grams of solute needed to prepare each of the following solutions:

1. How many grams of glucose are needed to prepare 400.0 mL of a 10.0% (w/v) solution?

2. How many grams of lidocaine hydrochloride are needed to prepare 50.0 g of a 2.0% (w/w) solution?

3. How many grams of KCl are needed to prepare 0.80 liters of a 0.15% (w/v) KCl solution?

4. How many grams of NaCl are needed to prepare 250 mL of a 1.0% (w/v) solution?

Answers: (1) 40.0 g (2) 1.0 g (3) 1.2 g (4) 2.5g

I. Use percent-concentration factors to calculate the volume (mL) of each solution that contains the amount of solute requested in each problem.

1. 2.00 g NaCl from a 1.00% (w/v) NaCl solution.

2. 25 g glucose from a 5.0% (w/v) glucose solution.

3. 1.5 g KCl from a 0.50% (w/v) KCl solution.

4. 75 g NaOH from a 25% (w/v) NaOH solution.

Answers: (1) 200 mL (2) 500 mL (3) 300 mL (4) 300 mL

MOLARITY (8.6)

Review: Molarity is a concentration term that describes the number of moles of solute dissolved in 1 L (1000 mL) of solution.

$$M = \frac{\text{moles}}{\text{L solution}}$$

J. Calculate the molarity of the following solutions:

1. 2.0 mole HCl in 1.0 liter

2. 10.0 mole glucose in 2.0 liters

3. 80.0 g NaOH in 4.0 liters
(Hint: Find moles NaOH.)

4. A 10.0% NaOH solution

Answers: (1) 2.0 M (2) 5.0 M (3) 0.50 M NaOH (4) 2.5 M

Review: Express the molarity (M) of a solution in the form of a conversion factor. For example, the factors that can be written for a 4 M solution are

$$\frac{4 \text{ moles}}{1 \text{ L}} \quad \text{and} \quad \frac{1 \text{ L}}{4 \text{ moles}}$$

Example: How many moles of NaOH are in 2 L of a 4.0 M NaOH solution?

Solution:

$$2 \text{ L} \quad \times \quad \frac{4.0 \text{ moles NaOH}}{1 \text{ L}} = 8 \text{ moles NaOH}$$

K. Calculate the quantity of solute in the following solutions:

1. How many moles of HCl are in 1.0 liter of a 4.0 M HCl solution?

2. How many moles of KOH are in 5.0 liters of a 2.0 M KOH solution?

3. How many grams of NaOH are needed to prepare 0.50 L of a 6.0 M NaOH solution? (Hint: Find moles of NaOH.)

4. How many moles of NaCl are in 200 mL of a 1.00 M NaCl solution.

Answers: (1) 4.0 moles (2) 10 moles (3) 120 g (4) 0.2 moles

L. Calculate the volume (L) needed of each solution to obtain the quantity of solute:

1. 10.0 moles of $Mg(OH)_2$ from a 2.0 M $Mg(OH)_2$ solution

2. 0.50 mole of glucose from a 5.0 M glucose solution

3. 0.10 mole of KI from a 1.0 M KI solution

4. 4.0 g NaOH from a 1.0 M NaOH solution

Answers: (1) 5.0 L (2) 0.10 L (3) 0.10 L (4) 0.10 L

DILUTIONS (8.7)

M. Solve the following dilution problems:

1. How much water is needed to make a 1:2 dilution of 100 mL of a 5.0 M KCl solution?

2. How many mL of water are needed to make a 1:5 dilution of 10 mL of a 15 % KCl solution?

3. You have 400 mL of a 5.0% NaOH solution which you need to dilute to a 1.0% NaOH solution.

   a. What is the dilution factor?

   b. What is the final volume of the solution?

   c. How much water must be added?

4. 120 mL of water is added to 40 mL of an 8.0% NaCl solution.

   a. What is the final volume?

   b. What is the dilution factor?

   c. What is the final concentration of the solution?

Answers: (1) 100 mL (2) 40 mL (3) a. 1:5 b. 2000 mL c. 1600 mL
(4) a. 160 mL b. 1:4 c. 2.0% NaCl

SOLUTIONS, COLLOIDS, AND SUSPENSIONS (8.8)

N. Identify each of the following as a solution, colloid, or suspension:

1.________ Gives the Tyndall effect.

2.________ Contains single atoms, ions, or small molecules less than 1 nm in size.

3.________ Settles out by gravity.

4.________ Retained by filters.

5.________ Cannot diffuse from a cellular membrane.

6.________ Aggregates of atoms, molecules or ions that are 1-100 nm in size.

7.________ Does not reflect a beam of light.

Answers: 1. colloid 2. solution 3. suspension 4. suspension 5. colloid 6. colloid 7. solution

OSMOSIS AND DIALYSIS (8.9)

Review: ***Osmosis*** is the movement of water across a semipermeable (osmotic) membrane from a solution of lesser concentration (more water) to one of greater concentration (less water).

Dialysis is the movement of water and small solution particles from an area where they have a higher concentration to an area where they have a lower concentration.

O. Complete:

In osmosis, the direction of solvent flow is from the (1)________ solvent concentration to the (2)__________solvent concentration. The movement of a solute throughout a solvent is called (3) ____________. Two sucrose solutions, a 5% and 10%, are separated by a semipermeable membrane. The (4) __________% solution has the greater osmotic pressure. Water will move from the (5)________% solution into the (6)_________% solution. The compartment containing the (7)__________% solution will increase in volume.

Answers: (1) higher (2) lower (3) diffusion (4) 10 (5) 5 (6) 10 (7) 10

P. A 2% and a 10% starch solution are separated by a semipermeable membrane.

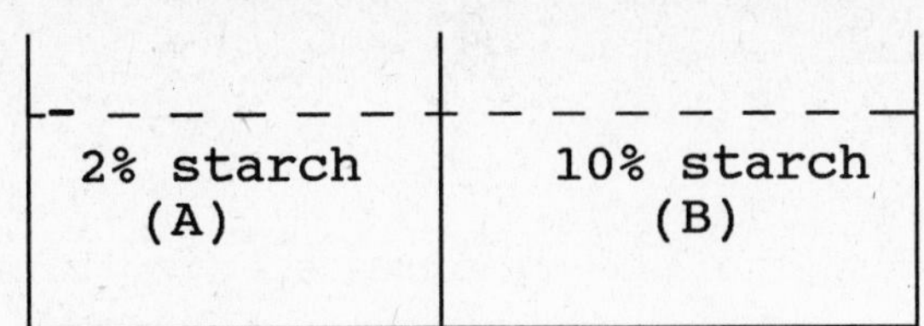

1. Water will flow from side _______ to side ________.

2. Compartment _____will increase in volume, and compartment _____ will decrease in volume.

3. The final concentration of the solutions in both compartments will be

___________.

Answers: 1. A, B 2. B, A 3. 6%

Q. Complete:

A (1)______% NaCl solutions and a (2)______% glucose solution are isotonic to the body fluids. A red blood cell placed in solutions (1) or (2) will not change in volume because these solutions are (3)______tonic to a red blood cell. When a red blood cell is placed in water, it (4)_____________ because water is (5)_______tonic to the red blood cell. A 20% glucose solution will cause a red blood cell to undergo (6)__________________ because the 20% glucose solu tion is (7)_____tonic to the red blood cell.

Answers: (1) 0.9 (2) 5 (3) iso (4) hemolyzes
(5) hypo (6) crenation (7) hyper

R. Indicate whether the following solutions are
A. hypotonic B. hypertonic C. isotonic

1.____ 5% glucose 2. ____ 3% NaCl

3.____ 2% glucose 4. ____ water

5.____ 0.9% NaCl 6. ____ 10% glucose

Answers: 1. C 2. B 3. A 4. A 5. C 6. B

S. Indicate whether the following solutions will cause a red blood cell to undergo
A. crenation or B. hemolysis or C. not change (stays the same).

1. ____ 10% NaCl   2. ____ 1% glucose

3. ____ 5% glucose   4. ____ 0.5% NaCl

5. ____ 10% glucose   6. ____ water

Answers: 1. A 2. B 3. C 4. B 5. A 6. B

T. A dialysis bag contains starch, glucose, NaCl, protein, and urea.

1. When the dialysis bag is placed in water, which of the components would you expect to dialyze through the bag?

   Why?

2. Which components will stay inside the dialysis bag?

   Why?

Answers 1. glucose, NaCl, urea; they are solution particles.
2. starch, protein; they are colloids and retained by semipermeable membranes.

SELF-EVALUATION TEST
SOLUTIONS

Indicate if the following are more soluble in
(A) water, a polar solvent or (B) benzene, a nonpolar solvent.

1. _____ $I_2(g)$, nonpolar

2. _____ NaBr(s), polar

3. _____ KI(s), polar

4. _____ C_6H_{12}, nonpolar

5. The solubility of NH_4Cl is 46 g in 100 g of water at 40°C. How much NH_4Cl can dissolve in 500 g of water at 40°C?
A. 46 g B. 92 g C. 100 g D. 184 g E. 230 g

For questions 6-10, indicate whether each statement describes a
A. solution B. colloid C. suspension

6. _____Contains single atoms, ions, or small molecules of solute.

7. _____Settles out upon standing.

8. _____Can be separated by filtering.

9. _____Can be separated by semipermeable membranes.

10. _____Shows the Tyndall effect.

11. A solution containing 8.0 g NaCl in 400 mL of solution has a weight-volume percent concentration of
A. 0.5% B. 1% C. 2% D. 4% E. 20%

12. A solution containing 0.60 g sucrose in 50.0 mL of solution has a percent concentration of
A. 0.3% B. 0.6% C. 1.2% D. 3% E. 4.5%

13. The amount of lactose needed to prepare 500 mL of a 3.0% solution for an infant's formula is
A. 0.30 g B. 1.5 g C. 3.0 g D. 15 g E. 30 g

14. The volume needed to obtain 40 g of glucose from a 5% solution is
A. 100 mL B. 200 mL C. 400 mL D. 500 mL E. 800 mL

15. The amount of NaCl needed to prepare 500 mL of a 4% NaCl solution is
A. 20 g B. 15 g C. 10 g D. 4 g E. 2 g

For questions 16-18, consider a 1:5 dilution of 200 mL of a 10% NaOH solution

16. The final volume after dilution is
A. 200 mL B. 300 mL C. 500 mL D. 800 mL E. 1000 mL

17. The amount of water added was
A. 200 mL B. 300 mL C. 500 mL D. 800 mL E. 1000 mL

18. The new concentration of the solution is
A. 50% B. 10% C. 5% D. 2% E. 1%

19. The number of moles of KOH in 400 mL of a 2 M KOH solution is
A. 80 B. 20 C. 8 D. 2 E. 0.8

20. The amount of NaOH (40.0 g/mole) needed to prepare 0.5 L of a 2 M NaOH is
A. 0.40 g B. 4.0 g C. 8.0 D. 40 g E. 80 g

For questions 21-23, consider a solution containing 20.0 g NaOH in a volume of 400 mL. (NaOH has a molar mass of 40.0 g/mole)

21. The % concentration of the solution is
A. 2% B. 4% C. 5% D. 10% E. 20%

22. The number of moles of NaOH in the sample is
A. 0.20 B. 0.40 C. 0.50 D. 2.0 E. 4.0

23. The molarity of the sample is
A. 0.25 M B. 0.5 M C. 1.0 M D. 1.25 M E. 1.5 M

24. The separation of colloids from solution particles by use of a membrane is called
A. osmosis B. dispersion C. dialysis D. hemolysis E. collodian

25. Any two solutions that have identical osmotic pressures are
A. hypotonic B. hypertonic C. isotonic D. isotopic E. blue

26. In osmosis, water flows
A. between solutions of equal concentrations.
B. from higher solute concentrations to lower solute concentrations.
C. from lower solute concentrations to higher solute concentrations.
D. from colloids to solutions of equal concentrations.
E. from lower solvent concentrations to higher solvent concentrations.

27. A normal red blood cell will shrink when placed in a solution that is
A. isotonic B. hypotonic C. hypertonic D. colloidal E. semitonic

28. A red blood cell undergoes hemolysis when placed in a solution that is
A. isotonic B. hypotonic C. hypertonic D. colloidal E. semitonic

29. An example of an isotonic solution is
A. 0.1% NaCl B. 0.9 % NaCl C. 5% NaCl
D. 10% glucose E. 15% glucose

30. Which of the following is hypertonic to red blood cells?
A. 0.5% NaCl B. 0.9% NaCl C. 1% glucose
D. 5% glucose E. 10% glucose

31. Which of the following is hypotonic to red blood cells?
A. 2.0% NaCl B. 0.9% NaCl C. 1% glucose
D. 5% glucose E. 10% glucose

For questions 32-36, select the correct term from the following:
A. isotonic B. hypertonic C. hypotonic D. osmosis E. dialysis

32. ____A solution with a higher osmotic pressure than the blood.

33. ____A solution of 10% NaCl surrounding a red blood cell.

34. ____A 1% glucose solution.

35. ____The cleansing process of the artificial kidney.

36. ____The flow of water up the stem of a plant.

37. In dialysis,
   A. dissolved salts and small molecules are separated from colloids.
   B. nothing but water passes through the membrane.
   C. only ions pass through a membrane.
   D. two kinds of colloids are separated.
   E. colloids are separated from suspensions.

38. A dialyzing membrane
   A. is a semipermeable membrane.
   B. allows only water and true solution particles to pass through.
   C. does not allow colloidal particles to pass through.
   D. All of the above
   E. None of the above

39. Which substance will remain inside a dialysis bag?
   A. water B. NaCl C. starch D. glucose E. Mg^{2+}

40. Waste removal in hemodialysis is based on:
   A. concentration gradients between the bloodstream and the dialysate
   B. a pH difference between the blood stream and the dialysate
   C. use of an osmotic membrane
   D. greater osmotic pressure in the bloodstream
   E. renal compensation.

ANSWERS FOR THE SELF-EVALUATION TEST

1. B	11. C	21. C	31. C
2. A	12. C	22. C	32. B
3. A	13. D	23. D	33. B
4. B	14. E	24. C	34. C
5. E	15. A	25. C	35. E
6. A	16. E	26. C	36. D
7. C	17. D	27. C	37. A
8. C	18. D	28. B	38. D
9. B	19. E	29. B	39. C
10. B	20. D	30. E	40. A

SCORING THE SELF-EVALUATION TEST

40 questions 2½ points each 100 points total

CHAPTER 9
ACIDS, BASES, AND SALTS

KEY CONCEPTS

1. An electrolyte is a substance that dissociates in water to form ions.
2. An insoluble salt is not soluble in water. When the ions of an insoluble salt are placed together, a precipitate (solid) forms.
3. A strong acid (or base) completely ionizes in water, while a weak acid (or base) only partially ionizes (less than 50%). An acid increases the hydronium concentration; a base increases the hydroxide concentration.
4. A small percentage of water molecules in a water sample ionize to form an equal number of hydronium and hydroxide ions.
5. The pH of a solution is related to the hydrogen-ion concentration of a solution.
6. Neutralization is a process whereby the ion in greater quantity is removed through the formation of water until the hydronium and hydroxide ions become equal. The solution produced is neutral.
7. A titration is a process whereby the concentration of a measured amount of acid is determined by measuring the volume of a base required to neutralize the acid.
8. A buffer that contains a weak acid or base along with its salt is capable of maintaining the pH of a solution.
9. The equivalent weight of an acid or base is that mass of substance that produces one mole of H^+ or OH^-. The equivalent weight of a salt is that mass of salt that provides one mole of electrical charge.

KEY WORDS

Use complete sentences to describe the following solution terms:

electrolyte

acid

base

pH

neutralization

buffer

LEARNING EXERCISES

ELECTROLYTES (9.1)

A. Write an equation for the solution reaction of each of the following salts:

1. LiCl ______
2. $Mg(NO_3)_2$ ______
3. Na_3PO_4 ______
4. K_2SO_4 ______
5. $MgCl_2$ ______

Answers :

1. $LiCl(s) \xrightarrow{H_2O} Li^+(aq) + Cl^-(aq)$
2. $Mg(NO_3)_2(s) \xrightarrow{H_2O} Mg^{2+}(aq) + 2\ NO_3^-(aq)$
3. $Na_3PO_4(s) \xrightarrow{H_2O} 3\ Na^+(aq) + PO_4^{3-}(aq)$
4. $K_2SO_4(s) \xrightarrow{H_2O} 2\ K^+(aq) + SO_4^{2-}(aq)$
5. $MgCl_2(s) \xrightarrow{H_2O} Mg^{2+}(aq) + 2\ Cl^-(aq)$

B. Indicate whether aqueous solutions of the following will contain ***ions, molecules,*** or ***both ions and molecules***. Write an equation for the formation of the solution:

1. glucose, $C_6H_{12}O_6$, a nonelectrolyte ______

______

2. NaOH, a strong electrolyte ______

______

3. K_2SO_4, a strong electrolyte ______________________

____

4. NH_4OH, a weak electrolyte ______________________

____

Answers:

1. molecules: $C_6H_{12}O_6(s) \xrightarrow{H_2O} C_6H_{12}O_6(aq)$
2. ions: $NaOH(s) \xrightarrow{H_2O} Na^+(aq) + OH^-(aq)$
3. ions: $K_2SO_4(s) \xrightarrow{H_2O} 2\,K^+(aq) + SO_4^{2-}(aq)$
4. ions and molecules: $NH_4OH \underset{}{\overset{H_2O}{\rightleftarrows}} NH_4^+(aq) + OH^-(aq)$

IONIC REACTIONS: FORMATION OF A SOLID (9.2)

C. Complete the following chart by writing S if the combination represents a *soluble* salt, and IS if the combination is an *insoluble* salt. Write the formulas of the insoluble salts.

	NO_3^-	Cl^-	SO_4^{2-}	S^{2-}
Ag^+				
K^+				
Ba^{2+}				
Pb^{2+}				

Answers:

S	IS, AgCl	IS, Ag_2SO_4	IS, Ag_2S
S	S	S	S
S	S	IS, $BaSO_4$	IS, BaS
S	IS, $PbCl_2$	IS, $PbSO_4$	IS, PbS

D. Write the ionic equation for the formation of a precipitate when the following solutions are mixed. If no precipitate would form, write *none*.

1. K_2CO_3 + $BaCl_2$ ______________________
2. $Pb(NO_3)_2$ + HCl ______________________
3. $AgNO_3$ + Na_2SO_4 ______________________
4. $Cu(NO_3)_2$ + Na_2S ______________________
5. Na_3PO_4 + $AgNO_3$ ______________________

Answers: 1. $Ba^{2+} + CO_3^{2-} \longrightarrow BaCO_3$ 2. $Pb^{2+} + 2\,Cl^- \longrightarrow PbCl_2$
3. $2\,Ag^+ + SO_4^{2-} \longrightarrow Ag_2SO_4$ 4. $Cu^{2+} + S^{2-} \longrightarrow CuS$
5. $3\,Ag^+ + PO_4^{3-} \longrightarrow Ag_3PO_4$

ACIDS AND BASES (9.3)

Review: Acids ionize in water to give H^+ (H_3O^+) and an anion. Bases dissociate in water to produce a cation and an OH^-. Strong acids and strong bases are ionized 100%; weak acids and weak bases are partially ionized.

E. Write equations for the ionization of the following acids in water:

1. HCl, a strong acid ______________________________

2. HF, a weak acid ______________________________

3. HNO_3, a strong acid ______________________________

Answers:

1. $HCl \xrightarrow{H_2O} H^+ + Cl^-$

2. $HF \underset{}{\overset{H_2O}{\rightleftarrows}} H^+ + F^-$

3. $HNO_3 \xrightarrow{H_2O} H^+ + NO_3^-$

F. Write equations for the ionization of the following bases in water:

1. NaOH, a strong base ______________________________

2. NH_4OH, a weak base ______________________________

3. $Mg(OH)_2$, a strong base ______________________________

Answers: 1. $NaOH \longrightarrow Na^+ + OH^-$ 2. $NH_4OH \underset{}{\overset{H_2O}{\rightleftarrows}} NH_4^+ + OH^-$
3. $Mg(OH)_2 \longrightarrow Mg^{2+} + 2OH^-$

G. Indicate if the following characteristics describe an (A) acid or (B) base.

1. ____Turns blue litmus red

2. ____Tastes sour

3. ____Contains more OH^- ions than H^+ ions

4. ____Neutralizes basic solutions

5. ____Tastes bitter

6. ____Turns red litmus blue

7. ____Contains more H^+ ions than OH^- ions

8. ____Neutralizes acidic solutions

Answers 1. A 2. A 3. B 4. A 5. B 6. B 7. A 8. B

IONIZATION OF WATER (9.4)

Review: The K_w of water can be used to determine the $[H^+]$ or the $[OH^-]$ of a solution when one of the two variables, $[H^+]$ or $[OH^-]$, is given.

$$K_w = [H^+][OH^-] = 1 \times 10^{-14}$$

Example: What is the $[H^+]$ in a solution that has $[OH^-] = 1 \times 10^{-9}$ M?

Solution: $$[H^+] = \frac{1 \times 10^{-14}}{1 \times 10^{-9}\ M} = 1 \times 10^{-5}\ M$$

H. Write the $[H^+]$ when the $[OH^-]$ has the following values:

1. $[OH^-] = 1 \times 10^{-12}$ M $[H^+] =$

2. $[OH^-] = 1 \times 10^{-5}$ M $[H^+] =$

3. $[OH^-] = 1 \times 10^{-7}$ M $[H^+] =$

4. $[OH^-] = 1.2 \times 10^{-4}$ M $[H^+] =$

5. $[OH^-] = 3.5 \times 10^{-8}$ M $[H^+] =$

Answers: (1) 1×10^{-2} M (2) 1×10^{-9} M (3) 1×10^{-7} M
(4) 8.3×10^{-11} M (5) 2.9×10^{-7} M

I. Use the K_w to determine the $[OH^-]$ when the $[H^+]$ has the following values:

1. $[H^+] = 1 \times 10^{-3}$ M $[OH^-] =$

2. $[H^+] = 1 \times 10^{-10}$ M $[OH^-] =$

3. $[H^+] = 1 \times 10^{-6}$ M $[OH^-] =$

4. $[H^+] = 2.8 \times 10^{-13}$ M $[OH^-] =$

5. $[H^+] = 8.6 \times 10^{-7}$ M $[OH^-] =$

Answers: (1) 1×10^{-11} M (2) $1 \text{ X } 10^{-4}$ M (3) $1 \text{ X } 10^{-8}$ M (4) 3.6×10^{-2} M (5) 1.2×10^{-8} M

THE pH OF SOLUTIONS (9.5)

J. State the most acidic pH in each group:

1. _____ 5 or 2
2. _____ 3, 7, or 10
3. _____ 12, 9 or 2
4. _____ 8. 7.5, 4.4, or 3.2
5. _____ 0.2, 1.5, or 2.3
6. _____ 5.5, 3.8, 11.2, 1.6

Answers: (1) 2 (2) 3 (3) 2 (4) 3.2 (5) 0.2 (6) 1.6

K. (a) Calculate the pH of the following solutions. Describe the solution.

	pH	*acidic, basic or neutral*
1. $[H^+] = 1 \times 10^{-8}$ M	________	________________
2. $[H^+] = 0.001$ M	________	________________
3. $[OH^-] = 1 \times 10^{-12}$ M	________	________________
4. $[OH^-] = 2 \times 10^{-5}$ M	________	________________
5. $[OH^-] = 1 \times 10^{-7}$ M	________	________________

Answers (1) 8, basic (2) 3, acidic (3) 2, acidic (4) 9.3, basic (5) 7, neutral

L. State whether the following pH values are acidic, basic or neutral:

1. ______ plasma, pH = 7.4
2. ______ soft drink, pH = 2.8
3. ______ maple syrup, pH = 6.8
4. ______ beans, pH = 5.0
5. ______ tomatoes, pH = 4.2
6. ______ lemon juice, pH = 2.2
7. ______ saliva, pH = 7.0
8. ______ eggs, pH = 7.8
9. ______ lime, pH = 12.4
10. ______ strawberries, pH = 3.0

Answers 1. basic 2. acidic 3. acidic 4. acidic 5. acidic
6. acidic 7. neutral 8. basic 9. basic 10. acidic

M. Complete the following table:

	[H^+]	*[OH^-]*	*pH*	*acidic, basic, neutral*
1.	________	1×10^{-12} M	______	________
2.	________	________	10	________
3.	5×10^{-8} M	________	______	________
4.	________	________	______	neutral
5.	________	________	1	________

Answers

	[H^+]	[OH^-]	pH	acidic, basic, neutral
1.	1×10^{-2} M	1×10^{-12} M	2	acidic
2.	1×10^{-10} M	1×10^{-4} M	10	basic
3.	5×10^{-8} M	2×10^{-7} M	7.3	basic
4.	1×10^{-7} M	1×10^{-7} M	7	neutral
5.	1×10^{-1} M	1×10^{-13} M	1	acidic

IONIC REACTIONS: ACID-BASE NEUTRALIZATION (9.6)

N. Write neutralization equations for the reactions between the following acids and bases:

1. hydrochloric acid and magnesium hydroxide
2. sulfuric acid and sodium hydroxide
3. nitric acid and potassium hydroxide
4. phosphoric acid and sodium hydroxide
5. sulfuric acid and ammonium hydroxide

Answers:

1. $2HCl + Mg(OH)_2 \longrightarrow MgCl_2 + 2H_2O$
2. $H_2SO_4 + 2NaOH \longrightarrow Na_2SO_4 + 2H_2O$
3. $HNO_3 + KOH \longrightarrow KNO_3 + H_2O$
4. $H_3PO_4 + 3NaOH \longrightarrow Na_3PO_4 + 3H_2O$
5. $H_2SO_4 + 2NH_4OH \longrightarrow (NH_4)_2SO_4 + 2H_2O$

ACID-BASE TITRATION (9.7)

O. Solve the following problems using the titration data given:

1. A 5.0 mL sample of HCl is placed in a flask. Titration required 15.0 mL of 0.20 M NaOH. What is the molarity of the HCl in the sample?

   $HCl + NaOH \longrightarrow NaCl + H_2O$

2. A 10.0 mL sample of H_3PO_4 is placed in a flask. Titration required 42.0 mL of 0.10 M NaOH. What is the molarity of the H_3PO_4?

   $H_3PO_4 + 3\ NaOH \longrightarrow Na_3PO_4 + 3\ H_2O$

Answers: (1) 0.60 M (2) 0.14 M

BUFFERS (9.8)

P. State whether each of the following represents a buffer system and why.

1. HCl + NaCl ______________________
2. K_2SO_4 ______________________
3. H_2CO_3 ______________________
4. $H_2CO_3 + NaHCO_3$ ______________________

Answers: 1. No. A strong acid is not a buffer.
2. No. A salt alone cannot act as a buffer.
3. No. A weak acid alone cannot act as a buffer.
4. Yes. A weak acid and its salt can act as a buffer system.

EQUIVALENTS AND NORMALITY (9.9)

Review: One equiv = The quantity of a substance that provides one mole H^+ or one mole OH^-, or one mole of positive or negative charge.

$$\text{Equiv weight} = \frac{\text{mass(g) of substance}}{\text{number of equiv}}$$

Q. Calculate the following:

1. number of equiv of HCl in 73 g HCl

2. number of equiv Ca^{2+} in 80.2 g of Ca^{2+}

3. grams of Cl^- in 0.40 eq of Cl^-

4. grams of K^+ in 4.00 meq K^+

Answers: (1) 2.0 equiv (2) 4.00 equiv (3) 14 g (4) 0.156 g

R. Calculate the normality (N) of the following solutions:

1. a KOH solution containing 56.1 g KOH in 2.00 L solution.

2. a H_2SO_4 solution containing 98.1 g H_2SO_4 in 0.500 L solution.

3. a $MgCl_2$ solution containing 95.3 g in 1.00 L solution.

Answers (1) 0.500 N (2) 4.00 N (3) 2.00 N

SELF-EVALUATION TEST
ACIDS, BASES, AND SALTS

1. Which of these aqueous solutions is a strong electrolyte?
A. NH_4OH B. H_2CO_3 C. sugar D. CH_3OH E. HCl

2. An acid is a compound which when placed in water yields this characteristic ion:

A. H^+ B. OH^- C. Na^+ D. Cl^- E. CO_3^{2-}

3. $MgCl_2$ would be classified as a(n)
A. acid B. base C. salt D. buffer E. nonelectrolyte

4. $Mg(OH)_2$ would be classified as a(n)
A. acid B. base C. salt D. buffer E. nonelectrolyte

5. When dissolved in water $Ca(NO_3)_2$ dissociates into

A. Ca^{2+} + $(NO_3)_2^{2-}$

B. Ca^+ + NO_3^-

C. Ca^{2+} + 2 NO_3^-

D. Ca^{2+} + 2 N^{5+} + 2 O_3^{6-}

E. $CaNO_3^+$ + NO_3^-

6. CH_3CH_2OH, ethyl alcohol, is a non-electrolyte. When placed in water it

A. dissociates completely.
B. dissociates partially.
C. does not dissociate.
D. makes the solution acidic.
E. makes the solution basic.

7. Acetic acid is a weak acid because
A. it forms a dilute acid solution.
B. it is isotonic.
C. it is less than 50% ionized in water.
D. it is a nonpolar molecule.
E. it can form a buffer.

8. A weak base when added to water
A. makes the solution slightly basic.
B. does not affect the pH.
C. dissociates completely.
D. does not dissociate.
E. makes the solution slightly acidic.

9. In the K_w expression for pure H_2O, the $[H^+]$ has the value

A. 1×10^{-7} M B. 1×10^{-1} M C. 1×10^{-14} M

D. 1×10^{-6} M E. 1×10^{-12} M

10. Of the following pH values, which is the most acidic pH?
A. 8 B. 5 C. 1.5 D. 3.2 E. 9

11. Of the following pH values, which is the most basic pH?
A. 10 B. 4 C. 2.2 D. 11 E. 9

For questions 12-14, consider a solution with a $[H^+] = 1 \times 10^{-11}$

12. The pH of the solution is

A. 1 B. 2 C. 3 D. 11 E. 14

13. The hydroxide ion concentration is

A. 1×10^{-1} M B. 1×10^{-3} M C. 1×10^{-4} M

D. 1×10^{-7} M E. 1×10^{-11} M

14. The solution is
A. acidic B. basic C. neutral D. a buffer E. neutralized

For questions 15-17, consider a solution with a $[OH^-] = 1 \times 10^{-5}$ M.

15. The hydrogen ion concentration of the solution is

A. 1×10^{-5} M B. 1×10^{-7} M C. 1×10^{-9} M

D. 1×10^{-10} M E. 1×10^{-14} M

16. The pH of the solution is
A. 2 B. 5 C. 9 D. 11 E. 14

17. The solution is
A. acidic B. basic C. neutral D. a buffer E. neutralized

18. Which is an equation for neutralization?
A. $CaCO_3 \longrightarrow CaO + CO_2$

B. $Na_2SO_4(s) \longrightarrow 2Na^+ + SO_4^{2-}$

C. $H_2SO_4 + 2\ NaOH \longrightarrow Na_2SO_4 + 2\ H_2O$

D. $Na_2O + SO_3 \longrightarrow Na_2SO_4$

E. $H_2CO_3 \longrightarrow CO_2 + H_2O$

19. What is the name given to components in the body that keep blood pH within its normal 7.35 to 7.45 range?
A. nutrients B. buffers C. metabolites D. regufluids E. neutralizers

20. What is true of a typical buffer system?
A. It maintains a pH of 7.0.
B. It contains a weak base.
C. It contains a salt.
D. It contains a strong acid and its salt.
E. It maintains the pH of a solution.

21. Which of the following would act as a buffer system?
A. HCl
B. Na_2CO_3
C. NaOH + $NaNO_3$
D. NH_4OH
E. Na_2CO_3 + H_2CO_3

22. What is the number of equivalents in 1 mole of Mg^{2+}?
A. 0.50 equiv B. 1 equiv C. 1.5 equiv D. 2 equiv E. 4 equiv

23. What is the equivalent weight of calcium, Ca^{2+}?
A. 10 g/equiv B. 20 g/equiv C. 40 g/equiv
D. 60 g/equiv E. 80 g/equiv

24. What is the normality of an HCl solution containing 4.0 equivalents HCl in 2.0 L of solution?
A. 0.50 N B. 1.0 N C. 2.0 N D. 4.0 N E. 8.0 N

25. In a titration, 5.0 equiv of NaOH will completely neutralize _____equiv of H_2SO_4?
A. 10 equiv B. 5.0 equiv C. 2.5 equiv
D. 2.0 equiv E. 1.0 equiv

ANSWERS TO THE SELF-EVALUATION TEST

1. E	6. C	11. D	16. C	21. E
2. A	7. C	12. D	17. B	22. D
3. C	8. A	13. B	18. C	23. B
4. B	9. A	14. B	19. B	24. C
5. C	10.C	15. C	20. E	25. B

SCORING THE SELF-EVALUATION TEST

25 Questions 4 points each 100 points total

CHAPTER 10
ALKANES: AN INTRODUCTION TO A STUDY OF ORGANIC COMPOUNDS

KEY CONCEPTS

1. Organic compounds contain carbon, have low melting and boiling points, burn easily, and most are not soluble in water.

2. All organic compounds contain carbon which always has four bonds. Organic compounds may be represented by structural formulas, condensed formulas and sometimes by molecular formulas.

3. Organic compounds are named according to the IUPAC names of the straight-chain alkanes.

4. The naming of organic compounds is based upon the IUPAC system; an organic name indicates the number of carbons in the longest chain and any groups attached to that chain.

5. When two or more structural formulas are possible for a molecular formula, they are called isomers.

6. A cycloalkane is named by the number of carbon atoms in the cyclic structure preceded by the prefix cyclo-.

7. A haloalkane is named in the IUPAC system by naming the halogen atom (chloro, bromo, etc.) attached to an alkane parent chain.

8. Combustion is a chemical reaction of a hydrocarbon with oxygen, and halogenation is a chemical reaction of an alkane with a halogen.

KEY WORDS

Using complete sentences, write a definition for each of the following key terms:

alkane

condensed structural formula

parent chain

isomer

combustion

LEARNING EXERCISES

PROPERTIES OF ORGANIC COMPOUNDS (10.1)

A. Identify the following characteristics as typical of organic (O) or inorganic (I) compounds:

1.____covalent bonds
2.____combustible
3.____high melting points
4.____ionic bonds
5.____contains carbon
6._____low boiling points
7._____soluble in water
8._____soluble in nonpolar solvents
9._____form long chains
10.____are not very combustible

Answers: 1. O 2. O 3. I 4. I 5. O 6. O 7. I 8. O 9. O 10. I

STRUCTURAL FORMULAS (10.2)

B. State the number of bonds each of these atoms typically forms in a molecule:

1. H___
2. C___
3. N___
4. S___
5. O___
6. Cl___
7. P____
8. Br___

Answers: (1) 1 (2) 4 (3) 3 (4) 2 (5) 2 (6) 1 (7) 3 (8) 1

C. Complete the following structural formulas by adding the correct number of hydrogen atoms.

1. C–C

2. C–C–C–C

3. C–C–C–C–C (with a C attached above the second C)

Answers:

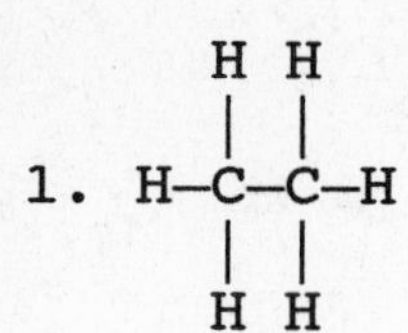

```
   H H H H
   | | | |
2. H-C-C-C-C-H
   | | | |
   H H H H
```

```
         H
         |
       H-C-H
     H   |  H H H
     |   |  | | |
3. H-C---C--C-C-C-H
     |   |  | | |
     H   H  H H H
```

D. Write the condensed formulas for the following structural formulas.

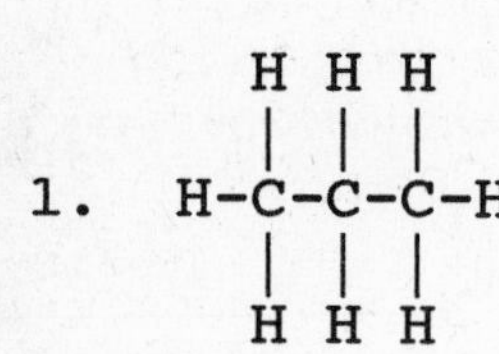

```
     H H H H H
     | | | | |
2. H-C-C-C-C-C-H
     | | | | |
     H H H H H
```

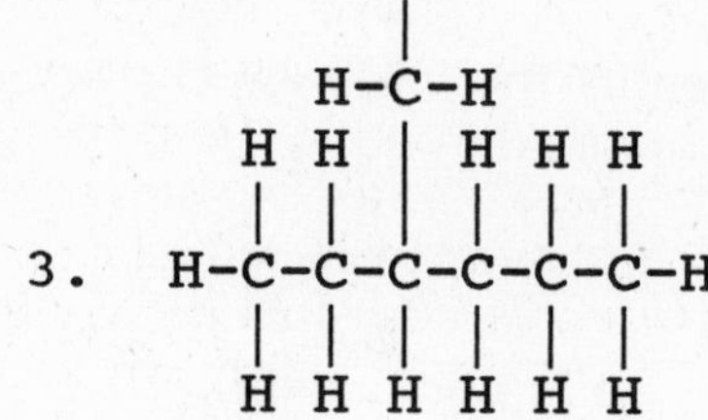

1. ______________________________

2. ______________________________

3. ______________________________

Answers 1. $CH_3CH_2CH_3$ 2. $CH_3CH_2CH_2CH_2CH_3$ 3. $CH_3CH_2\underset{}{\overset{CH_3}{\overset{|}{C}}}HCH_2CH_2CH_3$

IUPAC NAMES FOR ALKANES (10.3)

Review: In the IUPAC naming system, prefixes indicate the number of carbon atoms in a chain. Prefixes include *meth* (1C), *eth* (2C), *prop* (3C), *but* (4C), *pent* (5C), *hex* (6C), etc. The ending (suffix) of the name indicate the type of compound. Alkanes are named with the suffix *ane*. For example, an alkane with a chain of five carbon atoms is named *pentane*.

E. Write the condensed structure and name for the straight-chain alkane of each of the following molecular formulas:

1. C_2H_6 ______________________________

2. C_3H_8 ______________________________

3. C_4H_{10} ______________________________

4. C_5H_{12} ______________________________

5. C_6H_{14} ______________________________

Answers: 1. CH_3CH_3, ethane 2. $CH_3CH_2CH_3$, propane
3. $CH_3CH_2CH_2CH_3$, butane 4. $CH_3CH_2CH_2CH_2CH_3$, pentane
5. $CH_3CH_2CH_2CH_2CH_2CH_3$, hexane

IUPAC RULES FOR NAMING BRANCHED-CHAIN ALKANES (10.4)

Review: When there are carbon branches or other atoms in an alkane, the longest carbon chain is named as the *parent chain*. The carbon branches are named as alkyl groups and halogen atoms as halo groups. The alkyl names of carbon side groups are derived by replacing the *ane* of the alkane name with *yl*. For example, CH_3- is named as methyl (from CH_4 methane), and CH_3CH_2- is named as ethyl (from CH_3CH_3 ethane).

To name a compound with a side chain, number the parent chain from the end nearest the side chain. Precede the name of a side chain with the number of carbon to which it is attached.

Example: What is the name of $CH_3—\underset{}{\overset{CH_3}{\overset{|}{C}}}H—CH_2—CH_3$?

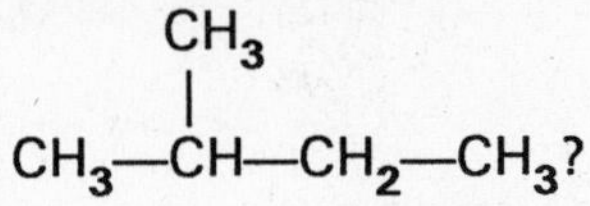

Solution: 2-methylbutane

F. Provide a correct IUPAC name for the following compounds:

1. 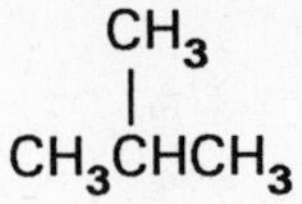 $CH_3\overset{CH_3}{\overset{|}{C}}HCH_3$ ____________________

2. 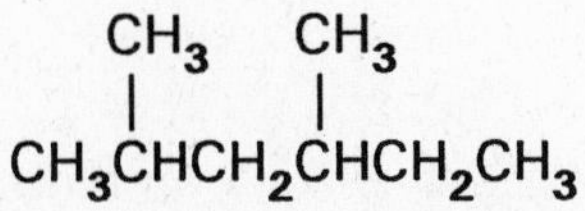 $CH_3\overset{CH_3}{\overset{|}{C}}HCH_2\overset{CH_3}{\overset{|}{C}}HCH_2CH_3$ ____________________

3. $CH_3\overset{CH_3}{\overset{|}{C}}HCH_2CH_2\overset{CH_3}{\overset{|}{C}}HCH_2CH_3$ ____________________

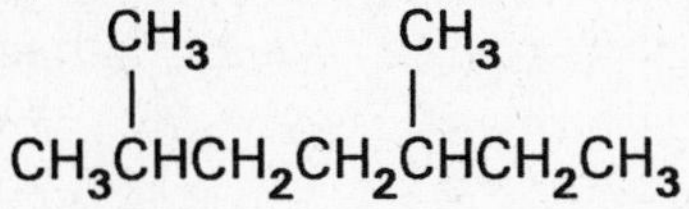

4. 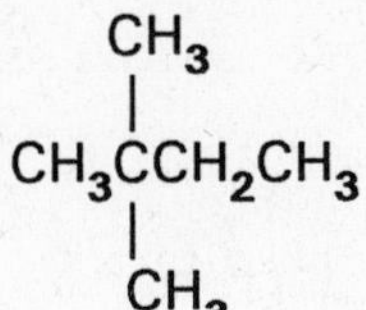 $CH_3\underset{\underset{CH_3}{|}}{\overset{CH_3}{\overset{|}{C}}}CH_2CH_3$ ____________________

5. 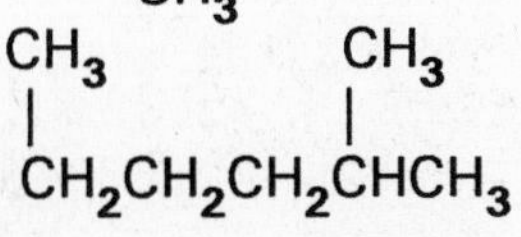 $\overset{CH_3}{\overset{|}{C}}H_2CH_2CH_2\overset{CH_3}{\overset{|}{C}}HCH_3$ ____________________

Answers: 1. methylpropane 2. 2,4-dimethylhexane 3. 2,5-dimethylheptane
4. 2,2-dimethylbutane 5. 2-methylhexane

G. Write the correct condensed formula for the following compounds:

1. hexane ______________________

2. methane ______________________

3. 2,4-dimethylpentane ______________________

4. propane ______________________

5. 2,3,4-trimethylpentane ______________________

Answers:

1. $CH_3CH_2CH_2CH_2CH_2CH_3$ 2. CH_4 3. $\begin{array}{c}CH_3\quad\;\; CH_3\\ |\qquad\quad |\\ CH_3CHCH_2CHCH_3\end{array}$

4. $CH_3CH_2CH_3$ 5. $\begin{array}{c}\quad CH_3\qquad\quad CH_3\\ \quad |\qquad\qquad |\\ CH_3\text{—}CH\text{—}CH\text{—}CH\text{—}CH_3\\ |\\ CH_3\end{array}$

ISOMERS (10.7)

***Review*:** Isomers are compounds that have the same number and kinds of atoms, but in different structural arrangements. For example, the following structural formulas have the same molecular formula, $C_2H_4Cl_2$, but different arrangements of atoms and different names.

$$\begin{array}{c}\quad Cl\\ \quad |\\ CH_3CH\text{—}Cl\end{array} \quad \text{and} \quad Cl\text{—}CH_2CH_2\text{—}Cl$$

1,1-dichloroethane 1,2-dichloroethane

H. *ESSAY*: Explain how it is possible to have two different compounds with the same molecular formula C_4H_{10}.

Answer to essay: Atoms with the same molecular formula can be arranged in different patterns to form different structures called isomers. For C_4H_{10}, the four carbon atoms may be arranged in a continuous straight chain, or they may be arranged as a chain of three carbon atoms with a one-carbon side group.

I. Write the condensed formulas and names for all the isomers with the following molecular formulas:

1. C_4H_{10} (two isomers)

2. C_5H_{12} (three isomers)

3. C_6H_{14} (five isomers)

Answers:

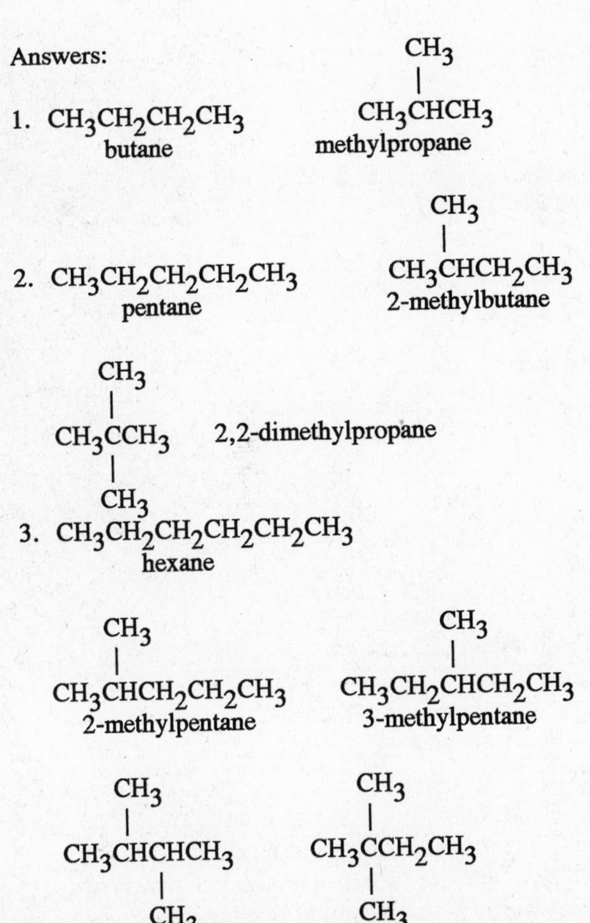

1. $CH_3CH_2CH_2CH_3$ butane; $CH_3CH(CH_3)CH_3$ methylpropane

2. $CH_3CH_2CH_2CH_2CH_3$ pentane; $CH_3CH(CH_3)CH_2CH_3$ 2-methylbutane; $CH_3C(CH_3)_2CH_3$ 2,2-dimethylpropane

3. $CH_3CH_2CH_2CH_2CH_2CH_3$ hexane; $CH_3CH(CH_3)CH_2CH_2CH_3$ 2-methylpentane; $CH_3CH_2CH(CH_3)CH_2CH_3$ 3-methylpentane; $CH_3CH(CH_3)CH(CH_3)CH_3$ 2,3-dimethylbutane; $CH_3C(CH_3)_2CH_2CH_3$ 2,2-dimethylbutane

CYCLOALKANES (10.6)

Review: For cycloalkanes, the prefix *cyclo* is used in front of the alkane name of the cyclic chain.

J. Write a correct IUPAC name for the following cycloalkanes

1. 1.____________________

2. 2.____________________

3. $-CH_3$ 3.____________________

4. $-CH_2CH_3$ 4.____________________

5. CH_3 CH_3 5.____________________

Answers: 1. cyclopropane 2. cyclohexane 3. methylcyclopentane 4. ethylcyclopentane 5. 1,1-dimethylcyclopropane

HALOALKANES (10.7)

K. Write a correct IUPAC (or common name) for the following

1. CH_3CH_2Cl ____________________

2. $CH_3CH_2CHClCH_3$ (Cl on C-3)

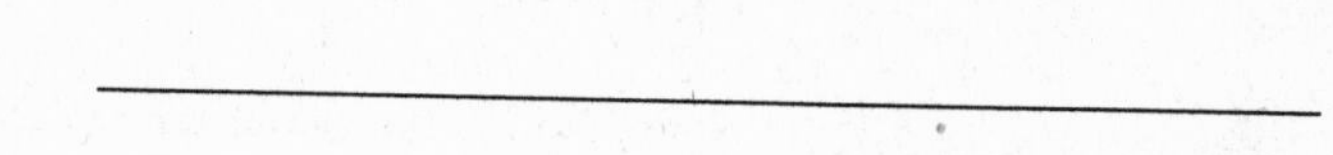

3. -Br ____________________

4. $CH_3CHClCH_2CHClCH_3$ ____________________

5. -F ____________________

Answers: 1. chloroethane (ethyl chloride) 2. 2-chlorobutane 3. bromocyclopentane (cyclopentylbromide) 4. 2,4-dichloropentane 5. fluorocyclopropane (cyclopropylfluoride)

L. Write the condensed formulas for the following haloalkanes:

1. ethyl chloride
2. bromoethane
3. 1-bromo-3-chlorocyclopentane
4. 1,1-dichlorocyclohexane
5. 2,2,3-trichlorobutane

Answers: 1. CH_3CH_2Cl 2. CH_3Br 3.

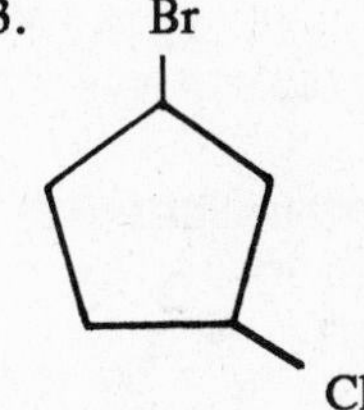

4. Cl Cl

5. $CH_3-\underset{Cl}{\overset{Cl}{C}}-\overset{Cl}{CH}-CH_3$

REACTIONS OF ALKANES (10.8)

Review: It is important to know what type of reaction a specific reagent brings about for a particular family of organic compounds. There are two reactions typical of alkanes: *combustion (oxidation)* which occurs with oxygen and heat, and a substitution of a hydrogen by a halogen called *halogenation*. This reaction occurs with a halogen in the presence of light or heat.

Examples:

Combustion: $CH_4 + 2\,O_2 \longrightarrow CO_2 + 2\,H_2O$

Halogenation $CH_4 + Br_2 + \text{light} \longrightarrow CH_3Br + HBr$

Note: The bromine atom can replace any one of the hydrogen atoms in the alkane.

M. Write a balanced equation for the complete combustion of the following:

1. propane ______________________________
2. hexane ______________________________
3. cyclopentane ______________________________

Answers:
1. $C_3H_8 + 5\,O_2 \longrightarrow 3\,CO_2 + 4\,H_2O$
2. $2\,C_6H_{14} + 19\,O_2 \longrightarrow 12\,CO_2 + 14\,H_2O$
3. $2\,C_5H_{10} + 15\,O_2 \longrightarrow 10\,CO_2 + 10\,H_2O$

N. Write the equation for the following halogenation reactions:

1. chlorination of methane in light

2. bromination of cyclopentane in light

3. chlorination of 2,2-dimethylpropane in light

4. bromination of cyclohexane in light

Answers:

1. $CH_4 + Cl_2 \xrightarrow{\text{light}} CH_3Cl + HCl$

2. cyclopentane $+ Br_2 \xrightarrow{\text{light}}$ cyclopentyl–Br $+ HBr$

3. $CH_3C(CH_3)_2CH_3 + Cl_2 \xrightarrow{\text{light}} CH_3C(CH_3)_2CH_2Cl + HCl$

4. cyclohexane $+ Br_2 \xrightarrow{\text{light}}$ cyclohexyl–Br $+ HBr$

SELF-EVALUATION TEST

ALKANES AN INTRODUCTION TO ORGANIC CHEMISTRY

Indicate whether the following characteristics are typical of (A) organic compounds or (B) inorganic compounds.

1. ___ higher melting points
2. ___ covalent bonds
3. ___ ionic bonds
4. ___ low boiling points
5. ___ fewer kinds of compounds
6. ___ soluble in water
7. ___ combustible
8. ___ soluble in nonpolar solvents

Match the name of the hydrocarbon with each structure.

9. $CH_3CH_2CH_3$
10. $CH_3CH_2CH_2CH_2CH_2CH_2CH_3$
11. CH_4
12. $CH_3CH_2CH_2CH_2CH_3$

A. methane
B. ethane
C. propane
D. pentane
E. heptane

Match the name of the hydrocarbon with each structure.

13. $CH_3CH_2CH_2CH_3$

14. $CH_3CH_2\underset{\displaystyle CH_3}{CH}CH_2\underset{\displaystyle CH_3}{CH}CH_3$ (methyl groups drawn above the two CH carbons)

15.

16. 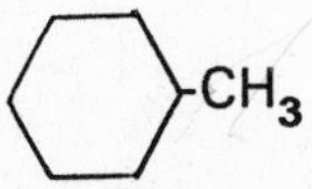 -CH_3

A. butane
B. methylcyclohexane
C. cyclopropane
D. 3,5-dimethylhexane
E. 2,4-dimethylhexane

Match the name of the hydrocarbon with each structure.

17.

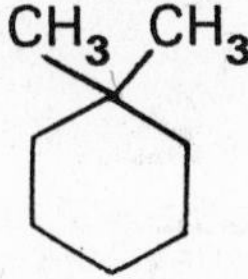

18. CH_3

19. CH_2CH_3

20.

A. methylcyclopentane
B. 1,2-dimethylcyclohexane
C. cyclobutane
D. 1,1-dimethylcyclohexane
E. ethylcyclopentane

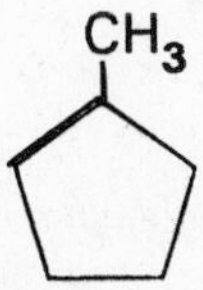

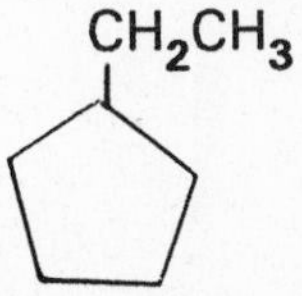

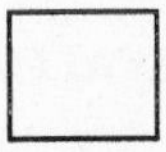

21. The correctly balanced equation for the complete combustion of ethane is

A. $C_2H_6 + O_2 \longrightarrow 2CO + 3H_2O$

B. $C_2H_6 + O_2 \longrightarrow CO_2 + H_2O$

C. $C_2H_6 + 2O_2 \longrightarrow 2CO_2 + 3H_2O$

D. $2C_2H_6 + 7O_2 \longrightarrow 4CO_2 + 6H_2O$

E. $2C_2H_6 + 4O_2 \longrightarrow 4CO_2 + 6H_2O$

22. In the presence of light, cyclohexane + Cl_2 gives

A. chlorobenzene + HCl
B. chlorocyclohexane + HCl
C. o-dichlorobenzene
D. dichlorocyclohexane + HCl
E. no reaction in light

Match the name of the compound with the correct structure.

23.
```
            Cl
            |
CH3CH2CH2CHCH2Cl
```

24.

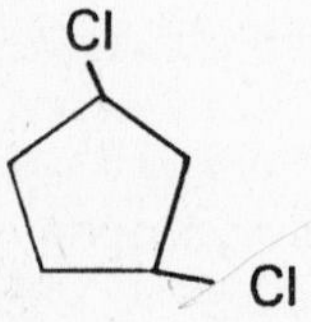

25.
```
  Cl    Cl
  |     |
CH3CHCH2CHCH3
```

A. 2,4-dichloropentane
B. 1,3-dichlorocyclopentane
C. 1,2-dichloropentane
D. 2,3-dichlorocyclopentane
E. 4,5-dichloropentane

ANSWERS TO THE SELF-EVALUATION TEST

1. B	6. B	11. A	16. B	21. D
2. A	7. A	12. D	17. D	22. B
3. B	8. A	13. A	18. A	23. C
4. A	9. C	14. E	19. E	24. B
5. B	10. E	15. C	20. C	25. A

SCORING THE SELF-EVALUATION TEST

25 questions 4 points each 100 points total

CHAPTER 11
UNSATURATED AND AROMATIC COMPOUNDS

KEY CONCEPTS

1. Alkenes contain a functional group consisting of a double bond, an area called an unsaturated site.

2. The double bond in an alkene gives the possibility of two isomers for the compound known as cis-trans isomers.

3. Alkynes contain a functional group consisting of a triple bond which is also an unsaturated site.

4. Alkenes and alkynes are reactive at their unsaturated sites where they can add hydrogen or halogens.

5. Aromatic hydrocarbons are organic compounds that contain a cyclic structure called benzene.

6. Aromatic compounds undergo substitution reactions that replace a hydrogen on the benzene ring with a halogen atom, nitro, or alkyl group.

KEY WORDS

Using complete sentences, write a description of each of the following key terms:

alkene

cis isomer

hydrogenation

halogenation

aromatic hydrocarbon

LEARNING EXERCISES

ALKENES (11.1)

Review: Alkenes are organic compounds that contain a double bond (C═C) as their functional group. Some examples of alkenes are the following:

$CH_3CH{=}CH_2$	$CH_2{=}CHCH_2CH_3$	$CH_3CH{=}CHCH_3$
propene	1-butene	2-butene

The IUPAC names of alkenes are derived by changing the *ane* ending of the parent alkane to *ene*. For example, the IUPAC name of $H_2C{=}CH_2$ is ethene. It has a common name of ethylene.

A. Write the IUPAC or common name for each of the following alkenes:

1. $CH_3CH{=}CH_2$

   1-ProPene

2. (cyclohexene ring structure)

   cyclohexene

3. $CH_3CH{=}C(CH_3)CH_3$ (CH₃ drawn above the third carbon)

   ____

4. $CH_3(H)C{=}C(H)CH_2CH_3$ (CH₃ and CH_2CH_3 on the same side, H below each carbon)

   cis-2-Pentene

5. $CH_2{=}CHCH(Cl)CH_2CH(CH_3)CH_3$ (Cl and CH_3 drawn above)

   5methyl-1-Hexene

6. $CH_3CH_2C({=}CH_2)CH_2CH_3$ (CH_2 double-bonded above the central carbon)

   ____

Answers: 1. propene; propylene 2. cyclohexene 3. 2-methyl-2-butene 4. cis-2-pentene 5. 3-chloro-5-methyl-1-hexene 6. 2-ethyl-1-butene

GEOMETRIC ISOMERS (11.2)

B. Write the cis-trans isomers of 2,3-dibromo-2-butene and name each.

Answers

$(CH_3)(Br)C{=}C(CH_3)(Br)$ — CH_3 groups on the same side, Br groups on the same side	$(CH_3)(Br)C{=}C(Br)(CH_3)$ — CH_3 groups on opposite sides
cis-2,3-dibromo-2-butene	trans-2,3-dibromo-2-butene

ALKYNES (11.3)

Review: The functional group in alkynes is a triple bond ($C \equiv C$). The IUPAC names of alkynes end in *yne*. Some examples of alkynes are seen below.

$HC \equiv CH$	$CH_3C \equiv CH$
ethyne	propyne
(acetylene)	(methylacetylene)

C. Write the IUPAC or common name of the following alkynes.

1. $HC \equiv CH$ ____________________
2. $CH_3C \equiv CH$
3. $CH_3CH_2C \equiv CH$ ____________________
4. $CH_3\overset{\overset{CH_3}{|}}{C}HC \equiv CCH_3$ ____________________

Answers: 1. ethyne; acetylene 2. propyne; methylacetylene 3. 1-butyne; ethylacetylene 4. 4-methyl-2-pentyne

CHEMICAL REACTIONS OF UNSATURATED HYDROCARBONS (11.4)

Review: The unsaturated hydrocarbons are very reactive at the site of the double or triple bond. Typical addition reactions include *hydrogenation* (addition of H_2/Pt), *halogenation* (addition of Br_2, Cl_2), *hydrohalogenation* (addition of HBr or HCl), and *hydration* (addition of water, H-OH in the presence of acid). The atoms from the reagent bond to the carbon atoms of the double (or triple) bond to form alkanes, haloalkanes, dihaloalkanes, and alcohols (-OH).

Examples:

Hydrogenation $CH_2{=}CH_2 + H_2 \xrightarrow{Pt} CH_3\text{-}CH_3$

Halogenation $CH_2{=}CH_2 + Br_2 \longrightarrow BrCH_2\text{-}CH_2Br$

Hydrohalogenation $CH_2{=}CH_2 + HCl \longrightarrow CH_3\text{-}CH_2Cl$

Hydration $CH_2{=}CH_2 + H\text{-}OH \xrightarrow{H^+} CH_3\text{-}CH_2OH$

D. Write the products of the following reactions:

1. $CH_3CH{=}CHCH_3 + HCl \longrightarrow$ $CH_3\overset{\overset{Cl}{|}}{C}H-CH_2CH_3$

2. $CH_3CH_2CH{=}CH_2 + H_2/Pt \longrightarrow$ $CH_3CH_2CH_2CH_3$

3. (cyclopentene) + H_2/Pt ⟶

4. $CH_3\overset{\displaystyle CH_3 \atop |}{C}{=}CH_2$ + HBr ⟶

5. $CH_3CH{=}CHCH_2CH_3$ + Cl_2 ⟶

6. $CH_3C{\equiv}CH$ + 2 H_2 ⟶

7. $CH_3C{\equiv}CCH_3$ + 2 Br_2 ⟶

8. $CH_3CH_2CH{=}\overset{\displaystyle CH_3 \atop |}{C}CH_3$ + HBr ⟶

Answers

1. $CH_3CH_2\overset{\displaystyle Cl \atop |}{C}HCH_3$ 2. $CH_3CH_2CH_2CH_3$ 3. (cyclopentane)

4. $CH_3\overset{\displaystyle CH_3 \atop |}{\underset{\displaystyle | \atop Br}{C}}CH_3$ 5. $CH_3\overset{\displaystyle Cl \atop |}{C}H\overset{\displaystyle Cl \atop |}{C}HCH_2CH_3$ 6. $CH_3CH_2CH_3$

7. $CH_3\overset{\displaystyle Br \atop |}{\underset{\displaystyle | \atop Br}{C}}{-}\overset{\displaystyle Br \atop |}{\underset{\displaystyle | \atop Br}{C}}{-}CH_3$ 8. $CH_3CH_2CH_2\overset{\displaystyle CH_3 \atop |}{\underset{\displaystyle | \atop Br}{C}}CH_3$

AROMATIC COMPOUNDS (11.5)

Review: In the IUPAC system, the benzene ring is numbered for two or more side groups. However, a commonly used system for naming benzene rings with two side groups is based on the use of the letters *o, m, p* rather than numbers to designate relative positions on the ring. The prefix *ortho- (o)* indicates a 1,2– attachment of side group; *meta- (m)* indicates that side chains are on the 1 and 3 carbon atoms; *para- (p)* is used for side groups on the 1 and 4 carbon atoms of the benzene ring.

E. Write the IUPAC (or common name) for each of the following.

1.

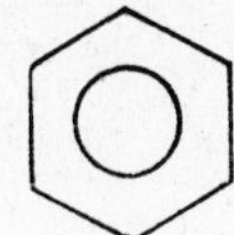

5.

Cl

Cl

2.

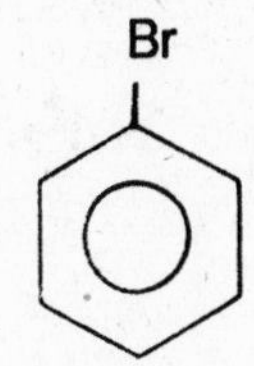

6.

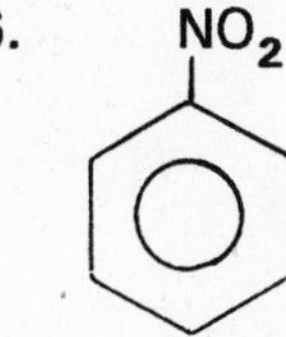

3.

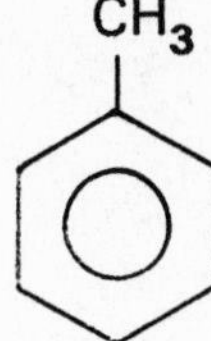

7.

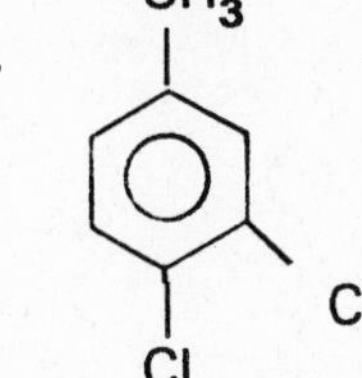

4.

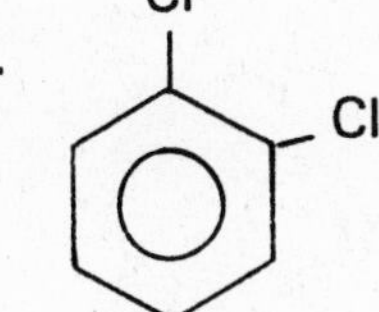

8.

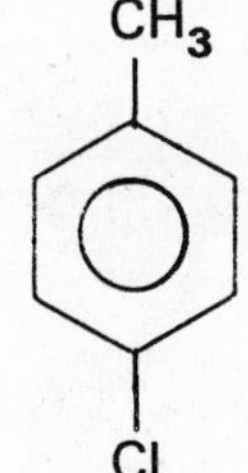

Answers: 1. benzene 2. bromobenzene 3. methylbenzene; toluene
4. 1,2-dichlorobenzene; o-dichlorobenzene
5. 1,3-dibromobenzene; m-dibromobenzene 6. nitrobenzene
7. 3,4-dichlorotoluene 8. p-chlorotoluene

CHEMICAL REACTIONS OF BENZENE (11.6)

Review: Benzene undergoes substitution reactions replacing a hydrogen atom with a halogen (Br_2 or Cl_2 and Fe), an alkyl group (CH_3-Cl with $AlCl_3$), a SO_3 (SO_3 with H_2SO_4), and a NO_2 (with HNO_3 and H_2SO_4).

F. Write the products of the following reactions of benzene.

1. + Cl_2 $\xrightarrow{Fe}$

2. Benzene + $CH_3Cl \xrightarrow{AlCl_3}$

3. Benzene + $SO_3 \xrightarrow{H_2SO_4}$

4. Benzene + $HNO_3 \xrightarrow{H_2SO_4}$

Answers: 1. Cl-benzene 2. CH_3-benzene 3. SO_3-benzene 4. NO_2-benzene

SELF-EVALUATION TEST UNSATURATED AND AROMATIC HYDROCARBONS

Questions 1, 2, 3, and 4 refer to $H_2C{=}CHCH_3$ (A) and cyclic H_2C—CH_2—CH_2 (B)

1. These compounds are
   A. aromatic B. alkanes C. isomers D. alkenes E. structural formulas

2. Compound (A) is a(n)
   A. alkane B. alkene C. cyclic hydrocarbon
   D. alkyne E. aromatic

3. Compound (B) is named
   A. propane B. propylene C. cyclobutane
   D. cyclopropane E. cycloproylene

4. Compound (A) is named
   A. propane B. propene C. 2-propene D. propyne E. 1-butene

In questions 5-8, match the name of the alkene with each structural formula.

5. $CH_2{=}CH_2$

6. 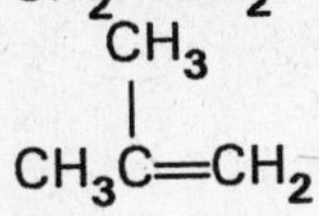 $CH_3\overset{CH_3}{\overset{|}{C}}{=}CH_2$

7. 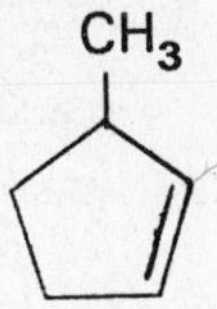 CH_3 (3-membered substituted cyclopentene ring structure)

8. 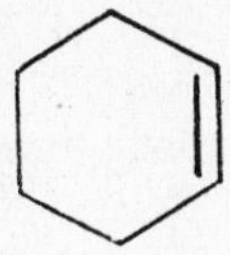 (cyclohexene ring structure)

A. cyclopentene
B. methylpropene
C. cyclohexene
D. ethene
E. 3-methylcyclopentene

9. The cis isomer of 2-butene is

A. $CH_2{=}CHCH_2CH_3$

B. $CH_3CH{=}CHCH_3$

C. 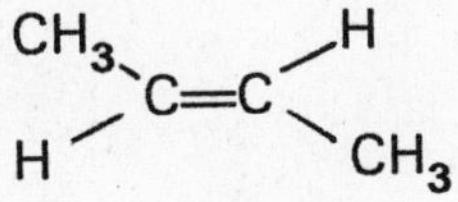 $\begin{matrix} CH_3 & & H \\ & C{=}C & \\ H & & CH_3 \end{matrix}$

D. 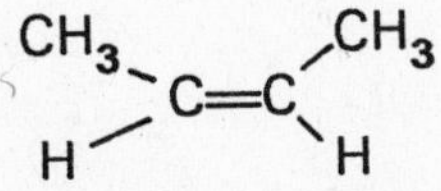 $\begin{matrix} CH_3 & & CH_3 \\ & C{=}C & \\ H & & H \end{matrix}$

E. 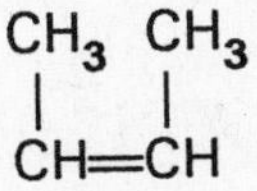 $\begin{matrix} CH_3 & CH_3 \\ | & | \\ CH{=}&CH \end{matrix}$

10. 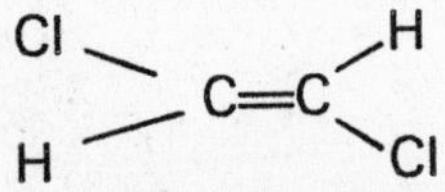 $\begin{matrix} Cl & & H \\ & C{=}C & \\ H & & Cl \end{matrix}$ is named

A. dichloroethene
B. cis-1,2-dichloroethene
C. trans-1,2-dichloroethene
D. cis-chloroethene
E. trans-chloroethene

11. Hydrogenation of $CH_3CH{=}CH_2$ gives

A. $3CO_2 + 6H_2$
B. $CH_3CH_2CH_3$
C. $CH_2{=}CHCH_3$
D. no reaction
E. $CH_3CH_2CH_2CH_3$

12. $CH_3CH{=}CH_2 + HBr \longrightarrow$

A. $CH_3CH_2CH_2Br$

B. no reaction

C. $CH_3CH_2CH_3$

D. $CH_3\overset{\overset{\displaystyle Br}{|}}{C}HCH_2Br$

E. $CH_3\overset{\overset{\displaystyle Br}{|}}{C}HCH_3$

13. Addition of bromine (Br_2) to ethene gives

A. CH_3CH_2Br B. $BrCH_2CH_2Br$ C. CH_3CHBr_2

D. CH_3CH_3 E. No reaction

14. Hydration of 2-butene gives

A. butane B. 1-butanol C. 2-butanol
D. cyclobutane E. cyclobutanol

15. What is the common name for the compound 1,3-dichlorobenzene?
A. *m*-dichlorobenzene
B. *o*-dichlorobenzene
C. *p*-dichlorobenzene
D. *x*-dichlorobenzene
E. *z*-dichlorobenzene

16. What is the common name of methylbenzene?
A. aniline B. phenol C. toluene D. xylene E. toluidine

17. What is the IUPAC name of $CH_3CH_2C{\equiv}CH$?

A. methylacetylene B. propyne C. propylene
D. 4-butyne E. 1-butyne

18. What is the product when benzene reacts with Br_2 in the presence of iron?
A. bromobenzene
B. *o*-dibromobenzene
C. *m*-dibromobenzene
D. *p*-dibromobenzene
E. No reaction

19. The reaction $CH_2{=}CH_2 + Cl_2 \longrightarrow ClCH_2CH_2Cl$ is called

A. hydrogenation of an alkene
B. halogenation of an alkene
C. hydrohalogenation of an alkene
D. hydration of an alkene
E. combustion

20. The reaction in problem 19 is
A. a hydrolysis reaction
B. an oxidation reaction
C. a substitution reaction
D. an addition reaction
E. a reduction reaction

21. The reaction $CH_3CH{=}CH_2 + HCl \longrightarrow CH_3\overset{Cl}{\overset{|}{C}}HCH_3$ is called
A. hydrogenation of an alkene
B. halogenation of an alkene
C. hydrohalogenation of an alkene
D. hydration of an alkene
E. combustion

For questions 22-25 identify the family for each compound as
A. alkane B. alkene C. alkyne D. aromatic

22. $CH_3CH{=}CH_2$

23. 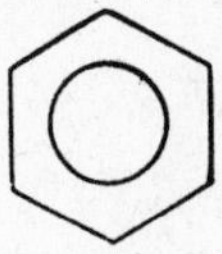

24. $CH_3CH_2\overset{CH_3}{\overset{|}{C}}HCH_2CH_3$

25. $CH_3CH_2C{\equiv}CH$

ANSWERS TO THE SELF-EVALUATION TEST

1. C	6. B	11. B	16. C	21. C
2. B	7. E	12. E	17. E	22. B
3. D	8. C	13. B	18. A	23. D
4. B	9. D	14. C	19. B	24. A
5. D	10. C	15. A	20. D	25. C

SCORING THE SELF-EVALUATION TEST

25 questions 4 points each 100 points total

CHAPTER 12
OXYGEN AND SULFUR IN ORGANIC COMPOUNDS

KEY CONCEPTS

1. An alcohol that contains the hydroxyl (-OH) group is named in the IUPAC system by replacing the *e* in the parent alkane name with *-ol* .

2. The hydrogen bonding of the -OH group makes alcohols more soluble in water than alkanes and gives them higher boiling points.

3. When benzene contains an -OH group, the compound is called a phenol.

4. The dehydration of an alcohol (removal of water) in the presence of an acid produces an alkene or an ether.

5. Ethers are compounds that contain an oxygen atom (-O-) between two alkyl groups and can be formed from two alcohols. An ether is commonly named by stating each alkyl name followed by the word *ether*.

6. Compounds containing a sulfur group –SH, –S–, or –S–S– are named as thiols, sulfides, or disulfides.

7. Aldehydes contain a carbonyl group at the end of the carbon chain; ketones contain the carbonyl group between two carbon atoms.

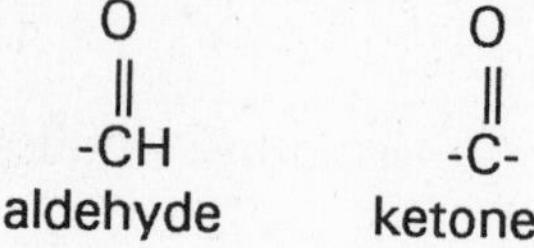

8. Alcohols can be classified as primary, secondary or tertiary alcohols. Primary alcohols oxidize (two hydrogens are removed) to form aldehydes; secondary alcohols oxidize to form ketones. There is no oxidation with tertiary alcohols. Aldehydes can be oxidized further to form carboxylic acids. Aldehydes and ketones can be reduced with hydrogen (H_2) to form alcohols.

KEY WORDS

Draw a structural formula containing the functional group for each of the following.

alcohol

ether

thiol

aldehyde

ketone

LEARNING EXERCISES

ALCOHOLS (12.1)

Review:: Alcohols contain the hydroxyl (-OH) functional group. In the IUPAC system alcohols are named by replacing the *ane* of the alkane name with *ol*. For example, CH_2OH is methan*ol*, and CH_3CH_2OH is ethan*ol*.

A. Give the correct IUPAC and common name (if any) for each alcohol.

1. CH_3CH_2OH

4. (cyclohexane ring with OH)

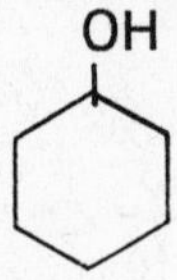

2. $CH_3CH_2CH_2OH$

5. $CH_3\overset{OH}{\overset{|}{C}}HCH_3$

3. $CH_3CH_2CH_2\overset{OH}{\overset{|}{C}}HCH_3$

6. (cyclopentane ring with CH_3 and OH on adjacent carbons)

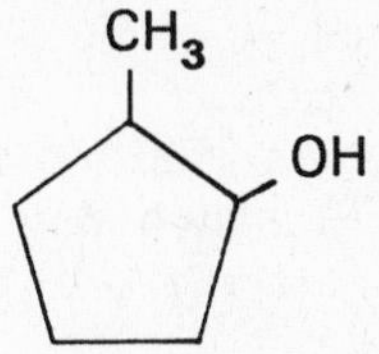

Answers: 1. ethanol; ethyl alcohol 2. 1-propanol; propyl alcohol 3. 2-pentanol 4. cyclohexanol 5. 2-propanol; isopropyl alcohol 6. 2-methylcyclopentanol

B. Write the correct condensed structure for the following alcohols.

1. 2-butanol
2. 2-chloro-1-propanol
3. 2,4-dimethyl-1-pentanol
4. cyclobutanol
5. 3-methylcyclopentanol

Answers:

1. $CH_3\overset{OH}{\overset{|}{C}}HCH_2CH_3$ 2. $CH_3\overset{Cl}{\overset{|}{C}}HCH_2OH$ 3. $CH_3\overset{CH_3}{\overset{|}{C}}HCH_2\overset{CH_3}{\overset{|}{C}}HCH_2OH$

4. (cyclobutane ring with OH) 5. (cyclopentane ring with OH and CH_3)

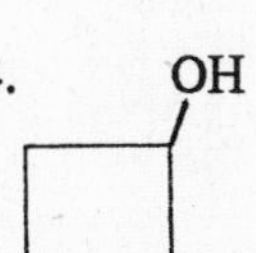

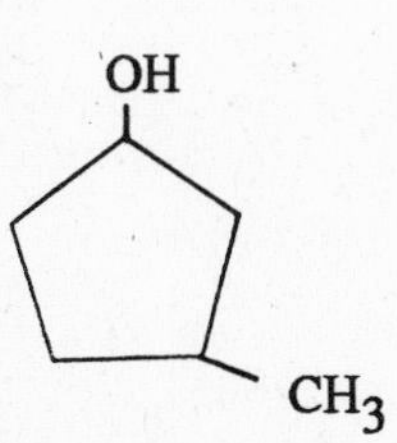

PHYSICAL PROPERTIES OF ALCOHOLS (11.2)

C. Circle the compound in each pair that is the more soluble in water.

1. CH_3CH_3 or CH_3CH_2OH
2. CH_3CH_2OH or $CH_3CH_2CH_2CH_2CH_2OH$
3. $CH_3CH_2CH_2OH$ or $CH_3CH_2CH_2CH_3$

Answers: 1. CH_3CH_2OH 2. CH_3CH_2OH 3. $CH_3CH_2CH_2OH$

PHENOLS (12.3)

D. Name the following phenols:

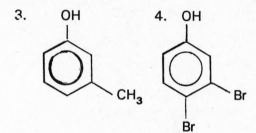

1. OH

2. OH

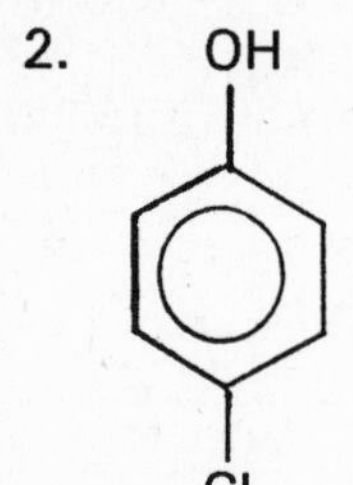

3. OH CH$_3$

4. OH Br Br

1. ____________________
2. ____________________
3. ____________________
4. ____________________

Answers: 1. phenol 2. p-chlorophenol 3. 3-methylphenol 4. 3,4-dibromophenol

DEHYDRATION OF ALCOHOLS (12.4)

Review: An alcohol is dehydrated (water removed) in the presence of an acid (H^+) and heat when the -OH group and an adjacent H- are removed to give an alkene and water.

Example: $CH_3CH_2OH \xrightarrow{H^+} H_2C{=}CH_2 + H_2O$

E. Write the structure of the product expected in the following dehydration reactions:

1. $CH_3CH_2CH_2CH_2OH \xrightarrow{H^+,\ heat}$

2. —OH $\xrightarrow{H^+,\ heat}$

3. $CH_3CH(OH)CH_3 \xrightarrow{H^+,\ heat}$

4. $CH_3CH(OH)CH_2CH_3 \xrightarrow{H^+,\ heat}$

Answers: 1. $CH_3CH_2CH{=}CH_2$ 2. (cyclohexene) 3. $CH_3CH{=}CH_2$
4. $CH_3CH{=}CHCH_3$ (major product)

ETHERS (12.5)

Review: Ethers contain the -O- functional group. In the IUPAC system, the O- and the shorter carbon chain are named as an *alkoxy* group attached to the longer alkane chain. However, simple ethers are most often named by the two alkyl groups followed by the word *ether*.

Example: $CH_3CH_2\text{-}O\text{-}CH_3$ IUPAC: methoxy ethane (common: ethyl methyl ether)

F. Write an IUPAC or common name for the following ethers.

1. CH_3OCH_3 ____________________

2. $CH_3CH_2OCH_2CH_3$ ____________________

3. $CH_3CH_2CH_2CH_2OCH_3$ ____________________

4. $CH_3OCH_2CH_3$ ____________________

5. $CH_3O\text{-}C_6H_5$ (benzene ring) ____________________

Answers: 1. methoxy methane; (di)methyl ether
2. ethoxy ethane: (di)ethyl ether
3. 1-methoxy butane; butyl methyl ether
4. methoxy ethane; ethyl methyl ether
5. methoxy benzene; methyl phenyl ether (anisole)

G. Write the structure of the ether formed in the following reactions.

1. $2\ CH_3CH_2OH \xrightarrow{H+,\ heat}$

2. $2\ CH_3OH \xrightarrow{H+,\ heat}$

Answers: 1. $CH_3CH_2OCH_2CH_3$ 2. CH_3OCH_3

THIOLS AND SULFIDES (12.6)

Review: Organic compounds with the functional group -SH are similar to alcohols and are called *thiols*. They are named by the alkane group followed by the word *thiol*. When sulfur is attached to two alkyl groups like an ether, the com pounds is called a *sulfide*. A compound containing a -S-S- group is called a *disulfide*.

Examples: CH_3SH methanethiol | CH_3-S-CH_3 (di)methyl sulfide | CH_3-S-S-CH_3 (di)methyl disulfide

H. Give the correct IUPAC or common names for the following thiols, sulfides or disulfides.

1. CH_3CH_2SH ____________________

2. $CH_3CH_2CH(CH_3)CH_2SH$ (CH_3 on the third carbon) ____________________

3. CH_3CH_2-S-S-CH_2CH_3 ____________________

4. $CH_3CH_2CH_2SH$ ____________________

5. CH_3CH_2-S-CH_3 ____________________

Answers: 1. ethanethiol (ethyl mercaptan) 2. 2-methyl-1-butanethiol 3. diethyl disulfide 4. 1-propanethiol (propyl mercaptan) 5. ethyl methyl sulfide

ALDEHYDES AND KETONES (12.7)

Review: Aldehydes and ketones contain the carbonyl group ($-\overset{O}{\overset{\|}{C}}-$). In aldehydes, the carbonyl group is always carbon 1. However, in ketones the carbonyl group occurs on a carbon atom within the carbon chain.

In the IUPAC system, ***aldehydes*** are named by replacing the *ane* in the alkane name by *al*. However, in the common naming system which is much in use, the commonly used prefixes appear: form (1C), acet (2C), propion (3C), butyr (4C), etc., followed by the word *aldehyde*. ***Ketones*** are named by replacing the *ane* in the alkane name by *one*. However, in the common naming system which is much in use, the alkyl groups on each side of the carbonyl group are named alphabetically followed by the word *ketone*.

Examples: $CH_3\overset{O}{\overset{\|}{C}}H$ | $CH_3\overset{O}{\overset{\|}{C}}CH_3$ CARBONYL GROUP

IUPAC name:	ethanal	propanone
common name:	acetaldehyde	dimethyl ketone

I. Classify the following compounds by their functional groups:

A. alcohol B. aldehyde C. ketone D. ether E. thiol

____1. $CH_3CH_2CH_2\overset{\overset{O}{\|}}{C}H$

____2. $CH_3CH_2CH_2OH$

____3. $CH_3CH_2\overset{\overset{O}{\|}}{C}CH_2CH_3$

____4. $CH_3CH_2OCH_3$

____5. $CH_3\overset{\overset{O}{\|}}{C}CH_2CH_3$

____6. $CH_3\overset{\overset{O}{\|}}{C}H$

____7. $CH_3CH_2\overset{\overset{SH}{|}}{C}HCH_3$

____8. $CH_3CH_2\overset{\overset{O}{\|}}{C}CH_3$

Answers: 1. aldehyde 2. alcohol 3. ketone 4. ether
5. ketone 6. aldehyde 7. thiol 8. ketone

J. Write the correct IUPAC (or common name) for the following aldehydes:

1. $CH_3\overset{\overset{O}{\|}}{C}H$ ________________________

2. $CH_3CH_2CH_2CH_2\overset{\overset{O}{\|}}{C}H$ ________________________

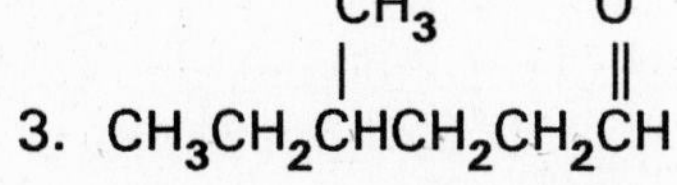

____

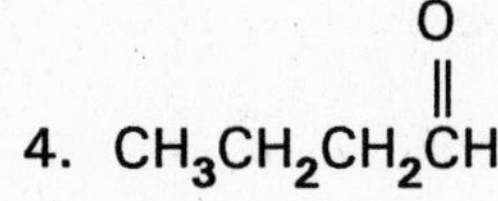

____

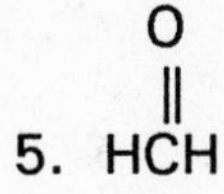

____

Answers: 1. ethanal; acetaldehyde
2. pentanal
3. 4-methylhexanal
4. butanal; butyraldehyde
5. methanal; formaldehyde

K. Write the correct IUPAC (or common name) for the following ketones.

1. $CH_3\overset{\overset{O}{\|}}{C}CH_3$ ____________________

2. $CH_3CH_2CH_2\overset{\overset{O}{\|}}{C}CH_3$ ____________________

3. $CH_3CH_2\overset{\overset{O}{\|}}{C}CH_2CH_3$ ____________________

4.  (cyclopentanone structure: O double-bonded to a five-membered ring) ____________________

5. 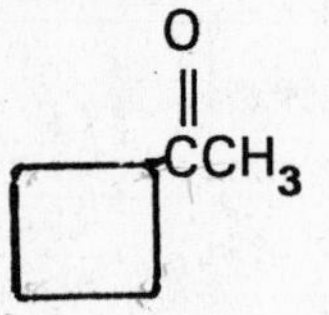 (cyclobutane ring bearing $\overset{\overset{O}{\|}}{C}CH_3$) ____________________

Answers: 1. propananone; dimethyl ketone, acetone 2. 2-pentanone; methyl propyl ketone 3. 3-pentanone; diethyl ketone 4. cyclopentanone 5. cyclobutyl methyl ketone

L. Write the correct condensed formulas for the following:

1. ethanal 2. α-methylbutyraldehyde 3. 2-chloropropanal

4. ethylmethylketone 5. 3-hexanone 6. benzaldehyde

Answers:

1. $CH_3\overset{\overset{O}{\|}}{C}H$ 2. $CH_3—CH_2—\overset{\overset{CH_3}{|}}{C}H—\overset{\overset{O}{\|}}{C}H$ 3. $CH3—\overset{\overset{Cl}{|}}{C}H—\overset{\overset{O}{\|}}{C}H$ 4. $CH_3\overset{\overset{O}{\|}}{C}CH_2CH_3$

5. $CH_3CH_2\overset{\overset{O}{\|}}{C}CH_2CH_2CH_3$ 6. 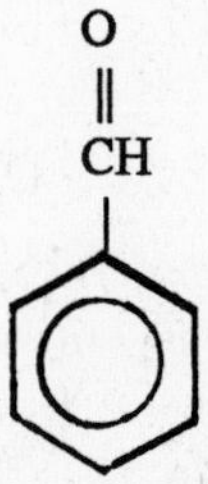 $\overset{\overset{O}{\|}}{C}H$ attached to a benzene ring

PREPARATION OF ALDEHYDES AND KETONES (12.8)

Review: Alcohols can be classified by the number of alkyl groups attached to the carbon bonded to the -OH group. 1° alcohols have one alkyl group; 2° alcohols contain two alkyl groups; 3° alcohols contains three alkyl groups.

Examples:

$CH_3—CH_2OH$ (1°)

$CH_3—CH(CH_3)—OH$ (2°)

$CH_3—C(CH_3)_2—OH$ (3°)

Review: When hydrogen atoms are removed by an oxidizing agent, primary and secondary alcohol oxidized. 1° alcohols oxidize to aldehydes, and 2° alcohols oxidize to ketones. Aldeh can oxidize further to carboxylic acids; 3° alcohols do not undergo oxidation.

Examples: CH_3CH_2OH (1° alcohol) $\xrightarrow{\text{oxidation}}$ $CH_3CH{=}O$ (aldehyde)

$CH_3CH(OH)CH_3$ (3° alcohol) $\xrightarrow{\text{oxidation}}$ $CH_3C(=O)CH_3$ (ketone)

M. Classify the following alcohols as primary (1°), secondary (2°), or tertiary (3°).

1.___ CH_3CH_2OH

2.___ 2-methyl-2-propanol

3.___ cyclohexane ring with —OH

4.___ $CH_3CH_2CH_2CH(OH)CH_2CH_3$

5.___ $CH_3C(OH)(CH_3)CH_2CH_3$

6.___ 2-hexanol

7.___ cyclobutanol

8.___ $CH_3CH(CH_3)CH_2OH$

9.___ methanol

10.___ cyclopentane ring with OH and CH_3 on the same carbon

Answers: (1) 1° (2) 3° (3) 2° (4) 2° (5) 3° (6) 2° (7) 2° (8) 1° (9) 1° (10) 3°

N. Write the oxidation product of each of the following alcohols:

1. $CH_3CH_2CH_2OH \xrightarrow{\text{oxidation}}$

2. $CH_3\overset{\overset{OH}{|}}{C}CH_2CH_3 \xrightarrow{\text{oxidation}}$

3. 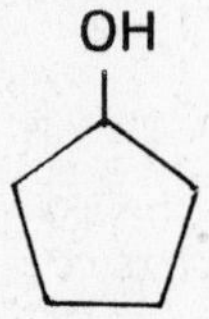 (cyclopentanol) $\xrightarrow{\text{oxidation}}$

4. $CH_3\overset{\overset{OH}{|}}{\underset{\underset{CH_3}{|}}{C}}CH_3 \xrightarrow{\text{oxidation}}$

5. methanol $\xrightarrow{\text{oxidation}}$

Answers: 1. $CH_3CH_2\overset{\overset{O}{\|}}{C}H$ 2. $CH_3\overset{\overset{O}{\|}}{C}CH_2CH_3$ 3. 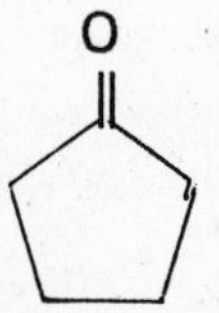 (cyclopentanone) 4. no reaction 5. $H\overset{\overset{O}{\|}}{C}H$

REACTIONS OF ALDEHYDES AND KETONES (12.9)

O. Write the reduction products for the following:

1. $CH_3\overset{\overset{O}{\|}}{C}CH_3 + H_2 \xrightarrow{Pt}$

2.  (cyclohexanone) $+ H_2 \xrightarrow{Pt}$

3. $CH_3CH_2CH_2\overset{\overset{O}{\|}}{C}H + H_2 \xrightarrow{Pt}$

4. 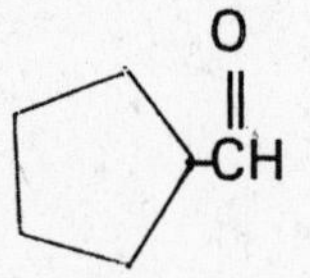 (cyclopentyl)$\overset{\overset{O}{\|}}{C}H + H_2 \xrightarrow{Pt}$

Answers: 1. $CH_3\overset{\overset{OH}{|}}{C}HCH_3$ 2. 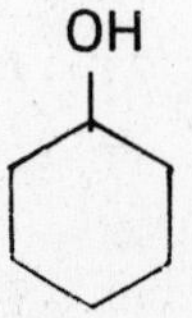 (cyclohexanol) 3. $CH_3CH_2CH_2CH_2OH$ 4. 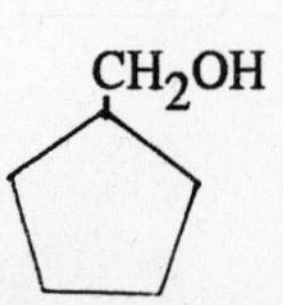 (cyclopentyl)CH_2OH

SELF-EVALUATION TEST

OXYGEN AND SULFUR IN ORGANIC COMPOUNDS

Match the names of the following compounds with their structures.

1. $CH_3\underset{}{\overset{OH}{|}}\!CHCH_3$ (OH on the central carbon)
2. $CH_3CH_2CH_2OH$
3. $CH_3OCH_2CH_3$
4. 
5. $CH_3CH_2OCH_2CH_3$

A. 1-propanol
B. cyclopropanol
C. 2-propanol
D. ethyl methyl ether
E. diethyl ether

6. The compound $CH_3\overset{O}{\overset{\|}{C}}CH_3$ is named
   A. 2-propanol B. propionaldehyde C. propanone
   D. propyl ether E. ethyl methyl ketone

7. Why are short-chain alcohols water soluble?
   A. They are nonpolar.
   B. They can hydrogen bond.
   C. They are organic.
   D. They are bases.
   E. They are acids.

8. Phenol is
   A. the alcohol of benzene
   B. the aldehyde of benzene
   C. the phenyl group of benzene
   D. the ketone of benzene
   E. cyclohexanol

9. $2\ CH_3CH_2OH \xrightarrow{H^+}$ ________
   A. an alkane B. an aldehyde C. a ketone
   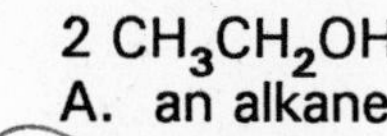
   D. an ether E. a phenol

10. The dehydration of cyclohexanol gives
    A. cyclohexane
    B. cyclohexene
    C. cyclohexyne
    D. benzene
    E. phenol

11. The formula of ethanethiol is
A. CH_3SH B. CH_3CH_2SH C. CH_3-S-CH_3
D. CH_3CH_2-S-CH_3 E. CH_3-S-S-CH_3

Match the following compounds with the names given.

12. $\overset{\displaystyle O}{\overset{\|}{HCH}}$

13. CH_3-O-CH_3

14. $CH_3\overset{\displaystyle O}{\overset{\|}{C}}CH_3$

15. $CH_3\overset{\displaystyle O}{\overset{\|}{C}}H$

A. dimethyl ether
B. acetaldehyde
C. methanal
D. dimethyl ketone
E. propanal

In questions 16, 17, 18, 19, and 20, classify each alcohol as
A. primary (1°) B. secondary (2°) C. tertiary (3°)

16. $CH_3\overset{\displaystyle OH}{\overset{|}{C}}HCH_3$

17. $CH_3\overset{\displaystyle OH}{\overset{|}{C}}HCH_2CH_2CH_2OH$

18. $CH_3\overset{\displaystyle CH_3}{\overset{|}{C}}HCH_2CH_2OH$

19. (cyclopentane ring)-OH

20. $CH_3CH_2\underset{\displaystyle CH_3}{\underset{|}{\overset{\displaystyle OH}{\overset{|}{C}}}}CH_3$

21. What is the name of the type of reaction that removes hydrogen from an organic compound?
    A. oxidation
    B. reduction
    C. hydrogenation
    D. halogenation
    E. hydration

Complete questions 22, 23, 24, and 25 using the terms below:

A. primary alcohols B. secondary alcohols C. aldehydes D. ketones E. carboxylic acids

22. Primary alcohols oxidize to form ________.

23. Secondary alcohols oxidize to form ________.

24. Aldehydes can be oxidized to form ________.

25. Ketones can be reduced to form ________.

ANSWERS TO SELF-EVALUATION TEST

1. C	6. C	11. B	16. B	21. A
2. A	7. B	12. C	17. B	22. C
3. D	8. A	13. A	18. A	23. D
4. B	9. D	14. D	19. B	24. E
5. E	10.B	15. B	20. C	25. B

SCORING THE SELF-EVALUATION TEST

25 questions 4 points each 100 points total

CHAPTER 13
CARBOXYLIC ACIDS, ESTERS, AND NITROGEN-CONTAINING COMPOUNDS

KEY CONCEPTS

1. Carboxylic acids contain the carboxyl functional group –COOH.

2. Carboxylic acids are weak acids which ionize slightly in water, have high melting and boiling points, and are neutralized by bases to give salts.

3. A carboxylic acid and an alcohol react in the presence of an acid to produce an ester. The ester is named from the alkyl portion of the alcohol followed by the name of the acid ending in *ate*.

4. Hydrolysis of an ester occurs in the presence of water to give an alcohol and carboxylic acid. When an ester is heated with a strong base such as NaOH or KOH, saponification occurs to give the salt of the acid (a soap).

5. Amines can be classified as primary, secondary, or tertiary. Amines are named by placing the ending ***amine*** after the name of the alkyl group(s).

6. A heterocyclic amine contains at least one nitrogen atom in a cyclic structure.

7. Amines act as weak bases in water by attracting the hydrogen from a water molecule to give an (alkyl)ammonium ion and a hydroxide ion. Amines are neutralized by acids to give salts of the amines.

8. Amides are formed by the reaction of a carboxylic acid and an amine. The OH of the carboxylic acid is replaced by the amine. Amides can be hydrolyzed by water to give a carboxylic acid and an amine.

KEY WORDS

Use a complete sentence to define each of the following terms.

carboxylic acid

ester

saponification

amine

amide

LEARNING EXERCISES

CARBOXYLIC ACIDS (13.1)

Review: Carboxylic acids contain the carboxyl (-COOH) functional group. They are named in the IUPAC system by replacing the *ane* ending with *oic acid.* However, simple acids usually are named by the common naming system using the prefixes we used for the aldehydes: ***form*** (1C), ***acet*** (2C), ***propion*** (4C), ***butyr*** (4C), etc. followed by ***ic acid.***

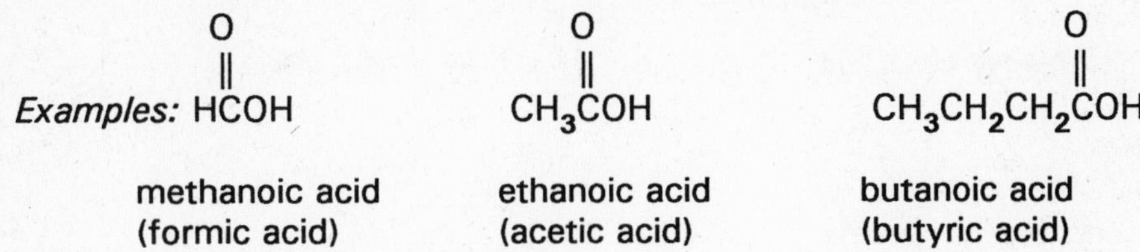

Examples:

methanoic acid (formic acid) — ethanoic acid (acetic acid) — butanoic acid (butyric acid)

A. Write the formulas for the following carboxylic acids:

1. acetic acid
2. 2-ketobutanoic acid
3. benzoic acid
4. ß-hydroxy propionic acid
5. formic acid

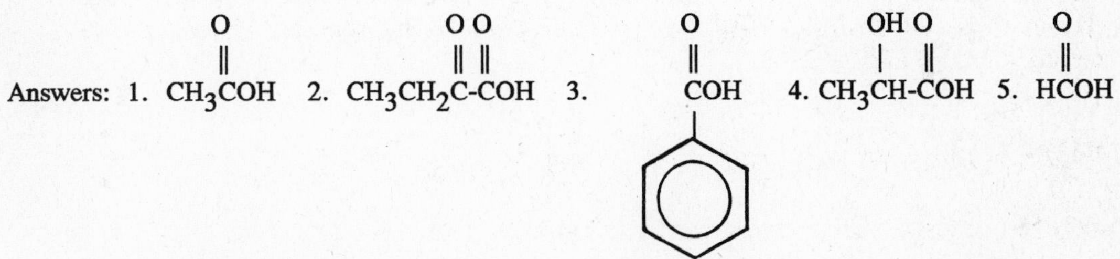

B. Write an equation for the oxidation of the appropriate aldehyde to produce the follow ing carboxylic acids:

1. propanoic acid

2. ß-methylbutyric acid

Answers: 1. $CH_3CH_2\overset{\overset{O}{\|}}{C}H \xrightarrow{\text{oxidation}} CH_3CH_2\overset{\overset{O}{\|}}{C}OH$

2. $CH_3\overset{\overset{CH_3}{|}}{C}HCH_2\overset{\overset{O}{\|}}{C}H \xrightarrow{\text{oxidation}} CH_3\overset{\overset{CH_3}{|}}{C}HCH_2\overset{\overset{O}{\|}}{C}OH$

SOME PROPERTIES OF CARBOXYLIC ACIDS (13.2)

Review: In water, carboxylic acids are weak acids which dissociate slightly to form an acidic solution of H^+ (H_3O^+) and a carboxylate ion. When a carboxylic acid reacts with a base, the products are a carboxylate salt and water.

C. Write the products for the following reactions of carboxylic acids.

1. $CH_3CH_2\overset{\overset{O}{\|}}{C}OH + H_2O \longrightarrow$

2. $CH_3CH_2\overset{\overset{O}{\|}}{C}OH + NaOH \longrightarrow$

3. formic acid + KOH $\longrightarrow$

4. benzoic acid + H_2O $\longrightarrow$

Answers: 1. $CH_3CH_2\overset{\overset{O}{\|}}{C}O^- + H_3O^+$ 2. $CH_3CH_2\overset{\overset{O}{\|}}{C}O^-Na^+ + H_2O$ 3. $H\overset{\overset{O}{\|}}{C}O^- K^+ + H_2O$

4. 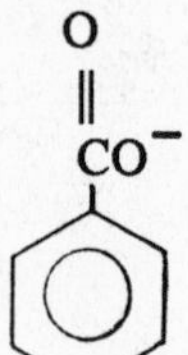 $C_6H_5\overset{\overset{O}{\|}}{C}O^- + H_3O^+$ (benzene ring bearing the $\overset{\overset{O}{\|}}{C}O^-$ group)

ESTERS (13.3)

Review: Esters contain the $-\overset{\overset{O}{\|}}{C}O-$ functional group. In the IUPAC system, they are named with the alkyl portion from the alcohol followed by the IUPAC name of the carboxylic acid which has the *oic acid* ending replaced by *oate*. The common names for esters are derived in the same way except that the common name of the acid is used and the *ic acid* ending replaced with *ate.*

Example: $CH_3\overset{O}{\overset{\|}{C}}\text{-}OCH_3$ methyl ethanoate; methyl acetate

D. Match the following carboxylic acids, salts of carboxylic acids and esters with their names:

1. $CH_3\overset{O}{\overset{\|}{C}}OH$
2. $C_6H_5\text{-}\overset{O}{\overset{\|}{C}}OH$
3. $CH_3CH_2CH_2\overset{O}{\overset{\|}{C}}O^-Na^+$
4. $CH_3CH_2CH_2\overset{O}{\overset{\|}{C}}OCH_3$
5. $CH_3CH_2CH_2\overset{O}{\overset{\|}{C}}OH$
6. $CH_3CH_2\overset{O}{\overset{\|}{C}}O^-Na^+$

A. ethyl acetate
B. butyric acid
C. sodium propanoate
D. sodium butyrate
E. methylbutyrate
F. ethylpropanoate
G. benzoic acid
H. acetic acid

Answers: 1. H 2. G 3. D 4. E 5. B 6. C

E. Write the ester product of the following esterification reactions.

Review: Esters are produced when a carboxylic acid and an alcohol react with an acid catalyst. A molecule of H_2O is removed as the esterification takes place.

1. $CH_3\overset{O}{\overset{\|}{C}}OH + CH_3OH \xrightarrow{H+}$
2. $C_6H_5\text{-}\overset{O}{\overset{\|}{C}}OH + CH_3OH \xrightarrow{H+}$
3. $H\overset{O}{\overset{\|}{C}}OH + CH_3CH_2OH \xrightarrow{H^+}$

Answers: 1. $CH_3\overset{O}{\overset{\|}{C}}OCH_3$ 2. $C_6H_5\text{-}\overset{O}{\overset{\|}{C}}OCH_3$ 3. $H\overset{O}{\overset{\|}{C}}OCH_2CH_3$

HYDROLYSIS AND SAPONIFICATION OF ESTERS (13.4)

Review: Esters are split by water to give a carboxylic acid and an alcohol. Their reactions with a strong base such as NaOH gives the salt of the carboxylic acid and an alcohol.

Examples:

$$\underset{\text{methyl acetate}}{CH_3\overset{O}{\overset{\|}{C}}\text{-}OCH_3} + H_2O \longrightarrow \underset{\text{acetic acid}}{CH_3\overset{O}{\overset{\|}{C}}OH} + \underset{\text{methyl alcohol}}{HOCH_3}$$

$$\underset{\text{methyl acetate}}{CH_3\overset{O}{\overset{\|}{C}}\text{-}OCH_3} + NaOH \longrightarrow \underset{\text{sodium acetate}}{CH_3\overset{O}{\overset{\|}{C}}O^-\ Na^+} + \underset{\text{methyl alcohol}}{HOCH_3}$$

F. Write the products of hydrolysis or saponification for the following esters:

1. $CH_3CH_2CH_2\overset{O}{\overset{\|}{C}}OCH_3 + H_2O \xrightarrow{H+}$

2. $CH_3\overset{O}{\overset{\|}{C}}OCH_3 + NaOH \longrightarrow$

3. $C_6H_5\text{-}\overset{O}{\overset{\|}{C}}OCH_2CH_3 + KOH \longrightarrow$

4. $C_6H_5\text{-}\overset{O}{\overset{\|}{C}}OCH_2CH_2CH_3 + H_2O \xrightarrow{H^+}$

Answers: 1. $CH_3CH_2CH_2\overset{O}{\overset{\|}{C}}OH + CH_3OH$ 2. $CH_3\overset{O}{\overset{\|}{C}}O^-Na^+ + CH_3OH$

3. $C_6H_5\overset{O}{\overset{\|}{C}}O^-K^+ + CH_3CH_2OH$ 4. $C_6H_5\overset{O}{\overset{\|}{C}}OH + CH_3CH_2CH_2OH$

G. Write structures for the following carboxylic acids, alcohols, salts of carboxylic acids, and esters:

1. potassium acetate
2. pentanoic acid
3. 1-butanol
4. 3-bromopropionic acid
5. ethylpropanoate
6. formic acid

Answers: 1. $CH_3\overset{\overset{O}{\|}}{C}O^-K^+$ 2. $CH_3CH_2CH_2CH_2\overset{\overset{O}{\|}}{C}OH$ 3. $CH_3CH_2CH_2CH_2OH$

4. $BrCH_2CH_2\overset{\overset{O}{\|}}{C}OH$ 5. $CH_3CH_2\overset{\overset{O}{\|}}{C}OCH_2CH_3$ 6. $H\overset{\overset{O}{\|}}{C}OH$

AMINES (13.5)

Review: Amines are nitrogen-containing compounds in which alkyl groups replace one, two or three hydrogen atoms of ammonia.

Examples:

NH_3	$CH_3{-}NH_2$	$CH_3{-}NH{-}CH_3$	$CH_3{-}\underset{\|}{\overset{CH_3}{\overset{\|}{N}}}{-}CH_3$
ammonia	methylamine (primary 1°)	dimethylamine (secondary 2°)	trimethylamine (tertiary 3°)

The number of groups attached to the N amines classifies amines as 1°, 2°, and 3°. Amines are named by giving the names of each alkyl group attached to the N atom. The amine of benzene is named aniline.

H. Write the names of the following amines and classify each as a primary (1°), secondary (2°), or tertiary (3°) amine:

	Name	Classification
1. $CH_3NHCH_2CH_3$	ethyl metlyAmine	________
2. (cyclohexane ring)$-NH_2$	cyHexoamm	________

3. $CH_3CH_2NH_2$ ____________________ ________

4. $CH_3CH_2—N(CH_3)—CH_3$ ____________________ ________

Answers: 1. ethylmethylamine, 2° 2. cyclohexanamine (cyclohexylamine), 2°
3. ethylamine, 1° 4. ethyldimethyamine, 3°

I. Write the structural formulas of the following amines:

1. 2-propanamine 2. N-methylaniline

Answers 1. $CH_3CH(NH_2)CH_3$ 2. C_6H_5–NH-CH_3

HETEROCYCLIC AMINES AND PHYSIOLOGICAL ACTIVITY (13.6)

J. Match the names of the following heterocyclic amines with the correct structure:
A. pyrrole B. pyridine C. pyrimidine D. purine E. pyrrolidine

1._____ 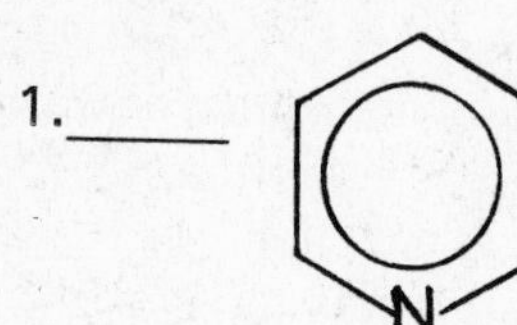

2._____ 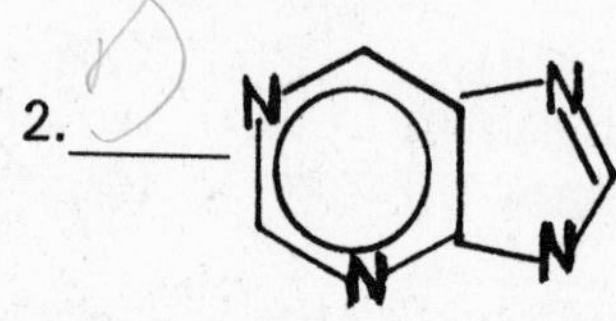

3._____ 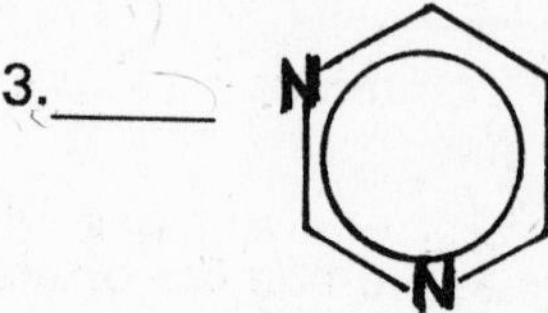

4._____ 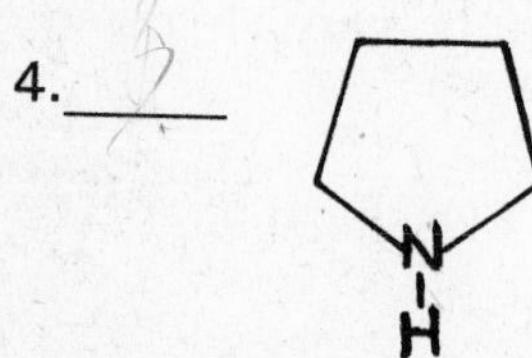

5._____ 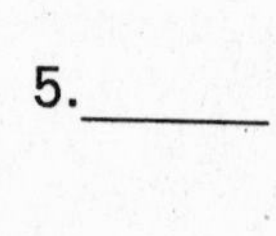 

Answers: 1. B 2. D 3. C 4. E 5. B

REACTIONS OF AMINES (13.7)

Review: In water, an amine (like ammonia) reacts as a weak base and accepts a proton from water.

Example: CH_3NH_2 + H-OH ⟶ $CH_3NH_3{}^+$ OH^-
methylamine → methylammonium hydroxide

Amines are neutralized by acids to form an ammonium salt.

Example: CH_3NH_2 + H-Cl ⟶ $CH_3NH_3{}^+$ Cl^-
methylamine → methylammonium chloride

K. Write the products of the following reactions of amines:

1. $CH_3CH_2NH_2$ + H_2O ⟶

2. $CH_3CH_2CH_2NH_2$ + HCl ⟶

3. $CH_3CH_2NHCH_3$ + HCl ⟶

4. (cyclohexane ring with NH_2) + H_2O ⟶

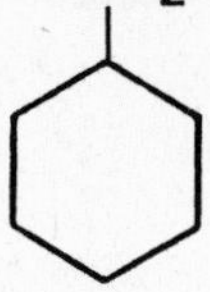

5. (cyclopentane ring with NH_2) + HBr ⟶

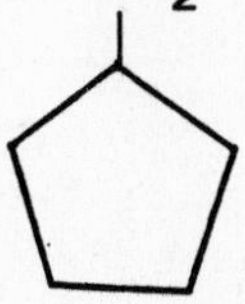

Answers 1. $CH_3CH_2NH_3{}^+$ OH^- 2. $CH_3CH_2CH_2NH_3{}^+$ Cl^- 3. $CH_3CH_2\overset{+}{N}H_2CH_3$ Cl^-

4. $NH_3{}^+$ OH^- (on cyclohexane ring) 5. $NH_3{}^+$ Br^- (on cyclopentane ring)

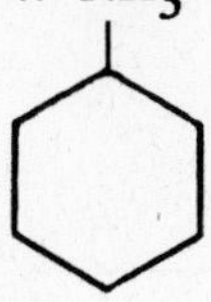

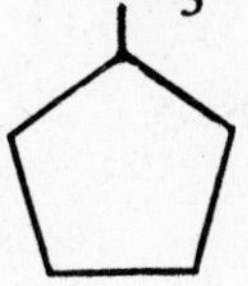

AMIDES (13.8)

Review: In the IUPAC system and common naming, amides are named by replacing the *-ic acid or -oic acid* ending with *amide*. When alkyl groups are attached to the N atom, they are listed as N-alkyl.

Examples: $CH_3\overset{\overset{\displaystyle O}{\|}}{C}\text{-}NH_2$ ethanamide (acetamide)

$CH_3\overset{\overset{\displaystyle O}{\|}}{C}\text{-}NHCH_3$ N-methylethanamide (N-methylacetamide)

Review: Amides can be formed when a carboxylic acid is heated with ammonia or an amine. The hydrolysis of an amide gives a carboxylic acid and an amine (or ammonia).

Example: $CH_3\overset{\overset{\displaystyle O}{\|}}{C}OH + NH_2CH_3 \xrightarrow{\text{heat}} CH_3\overset{\overset{\displaystyle O}{\|}}{C}\text{—}NHCH_3 + H_2O$

L. Name the following amides:

1. $CH_3CH_2\overset{\overset{\displaystyle O}{\|}}{C}NH_2$ ________________________

2. 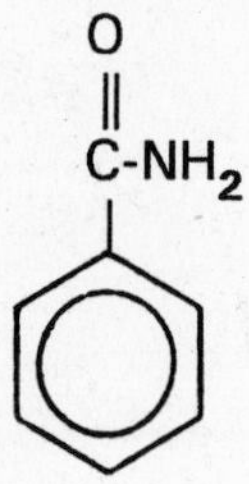 $C_6H_5\text{-}\overset{\overset{\displaystyle O}{\|}}{C}\text{-}NH_2$ (C-NH_2 attached to a benzene ring) ________________________

3. $CH_3CH_2CH_2CH_2\overset{\overset{\displaystyle O}{\|}}{C}NCH_3$ ________________________

4. $CH_3\overset{\overset{\displaystyle O}{\|}}{C}NHCH_2CH_3$ ________________________

5. 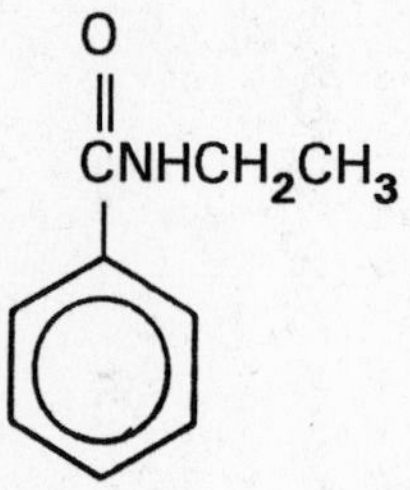 $C_6H_5\text{-}\overset{\overset{\displaystyle O}{\|}}{C}NHCH_2CH_3$ ($\overset{\overset{\displaystyle O}{\|}}{C}NHCH_2CH_3$ attached to a benzene ring) ________________________

Answers 1. propanamide (propionamide) 2. benzamide 3. N-methylpentanamide 4. N-ethylethanamide (N-ethyl acetamide) 5. N-ethylbenzamide

M. Write the structural formulas(s) of the product(s) formed in each of the following reactions:

1. $CH_3CH_2\overset{O}{\overset{\|}{C}}OH + NH_3 \xrightarrow{\text{heat}}$

2. $CH_3\overset{O}{\overset{\|}{C}}\text{-}NHCH_3 + H_2O \xrightarrow{\text{heat}}$

3. $C_6H_5\text{-}\overset{O}{\overset{\|}{C}}OH + CH_3NH_2 \xrightarrow{\text{heat}}$

4. $CH_3\overset{O}{\overset{\|}{C}}OH + \underset{NHCH_3}{\overset{CH_3}{|}} \xrightarrow{\text{heat}}$

Answers: 1. $CH_3CH_2\overset{O}{\overset{\|}{C}}\text{-}NH_2$

2. $CH_3\overset{O}{\overset{\|}{C}}OH + NH_2CH_3$

3. $C_6H_5\text{-}\overset{O}{\overset{\|}{C}}NHCH_3$

4. $CH_3\text{—}\overset{O}{\overset{\|}{C}}\text{-}\overset{CH_3}{\overset{|}{N}}\text{-}CH_3$

SELF-EVALUATION TEST
CARBOXYLIC ACIDS, ESTERS, AND NITROGEN-CONTAINING COMPOUNDS

Match the names of the following compounds with their structures:

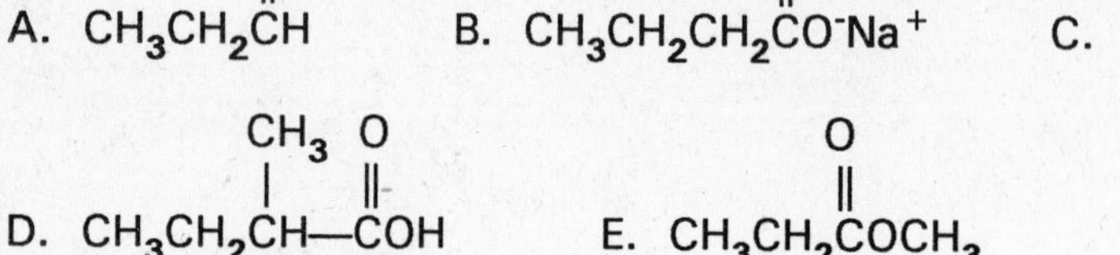

C. (cyclohexanone ring with C=O)

D. $CH_3CH_2CH(CH_3)—C(=O)OH$ E. $CH_3CH_2C(=O)OCH_3$

1.____ α-methylbutyric acid   2.____ methylpropanoate

3.____ sodium butanoate   4.____ propanal

5.____ cyclohexanone

6. A secondary alcohol can be oxidized to give a(n)
   A. aldehyde B. ketone C. carboxylic acid D. ester E. No reaction

7. What is the product when an aldehyde is reduced by hydrogen?
   A. 1° alcohol B. 3° alcohol C. 3° alcohol
   D. ketone E. no reaction

8. Carboxylic acids are water soluble due to their
   A. non-polar nature.
   B. ionic bonds.
   C. ability to lower pH.
   D. ability to hydrogen bond.
   E. high melting points.

Classify the amines in questions 9, 10, 11, and 12 as

A. primary amines B. secondary amines C. tertiary amines

____9. $CH_3CH(CH_3)NH_2$

____10. $CH_3CH_2N(CH_3)CH_3$

____11. $CH_3CH_2CH(NH_2)CH_2CH_3$

____12. $CH_3\text{-}N(CH_3)\text{-}CH_2CH_3$

Question 13, 14, 15, 16, 17 and 18 refer to the following reactions:

A. $CH_3\overset{O}{\overset{\|}{C}}OH + CH_3OH \xrightarrow{H^+} CH_3\overset{O}{\overset{\|}{C}}OCH_3 + H_2O$

B. $CH_3\overset{O}{\overset{\|}{C}}OH + NaOH \longrightarrow CH_3\overset{O}{\overset{\|}{C}}O^- \, Na^+ + H_2O$

C. $CH_3\overset{O}{\overset{\|}{C}}OCH_3 + H_2O \xrightarrow{H^+} CH_3\overset{O}{\overset{\|}{C}}OH + CH_3OH$

D. $CH_3\overset{O}{\overset{\|}{C}}OCH_3 + NaOH \longrightarrow CH_3\overset{O}{\overset{\|}{C}}O^-Na^+ + CH_3OH$

E. $CH_3\overset{O}{\overset{\|}{C}}OH + NH_3 \xrightarrow{heat} CH_3\overset{O}{\overset{\|}{C}}\text{-}NH_2 + H_2O$

13. Ester hydrolysis is reaction ______.

14. Amide synthesis is reaction ______.

15. Saponification is reaction ______.

16. Esterification is reaction ______.

17. What is the name of the organic product in reaction E?
A. methyl acetate B. benzamide C. acetamide
D. ethylamine E. ammonium acetate

18. What is the name of the product of reaction A?
A. methyl acetate B. acetic acid C. methyl alcohol
D. acetaldehyde E. ethyl methanoate

Match the amines and amides in questions 19, 20, 21, and 22 with the following names.
A. ethyl dimethyl amine B. butanamide C. N-methylacetamide
D. benzamide E. N-ethylbutyramide

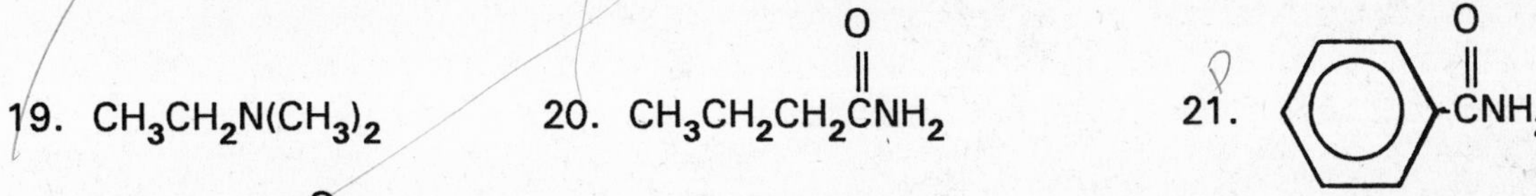

19. $CH_3CH_2N(CH_3)_2$

20. $CH_3CH_2CH_2\overset{O}{\overset{\|}{C}}NH_2$

21. $C_6H_5\text{-}\overset{O}{\overset{\|}{C}}NH_2$

22. $CH_3CH_2CH_2\overset{O}{\overset{\|}{C}}NHCH_2CH_3$

Match the products for the following reactions:

A. $CH_3CH_2\overset{\overset{O}{\|}}{C}OH + NH_3$

B. $CH_3CH_2NH_3^+ \ Cl^-$

C. $CH_3CH_2CH_2NH_3^+ \ OH^-$

23._____ ionization of 1-propanamine in water

24._____ hydrolysis of propanamide

25._____ reaction of ethanamine and hydrochloric acid

ANSWERS TO THE SELF-EVALUATION TEST

1. D	6. B	11. A	16. A	21. D
2. E	7. A	12. C	17. C	22. E
3. B	8. D	13. C	18. A	23. C
4. A	9. B	14. E	19. A	24. A
5. C	10. C	15. D	20. B	25. B

SCORING THE SELF-EVALUATION TEST

25 questions 4 points each 100 points total

CHAPTER 14
CARBOHYDRATES

KEY CONCEPTS

1. A monosaccharide is an organic compound that contains an aldehyde or ketone group. It is classified as an aldo- or ketotriose, tetrose, pentose, or hexose.

2. A chiral molecule contains four different atoms or groups attached to one carbon atom. A chiral molecule has a mirror image that is not superimposable.

3. The Fischer projections of chiral sugars placed the groups that project in front on the horizontal line, and the groups that project back on the vertical lines. In D-isomers, the -OH on the chiral carbon farthest from the carbonyl group is on the right side. It is on the left side in L isomers.

4. The Haworth structure of a saccharide represents the more prevalent cyclic structure linked by a hemiacetal bond between an -OH and the carbonyl carbon.

5. During the mutarotation of the α and ß cyclic structures of monosaccharides, the open chain provides an aldehyde group that can be oxidized by reagents such as Benedict's reagent. This gives the monosaccharides the name of reducing sugars. The reaction of an alcohol group with the hemiacetal bond of a sugar gives a glycosidic bond, the bond that links monosaccharides in di- and polysaccharides.

6. Dietary disaccharides—maltose, sucrose, and lactose are composed of a glucose unit and one other monosaccharide.

7. Polysaccharides—cellulose, starch, and glycogen—are polymers of glucose.

8. The types of carbohydrates can be identified by testing with Benedict's reagent, fermentation test, and iodine.

KEY WORDS

Using complete sentences, describe the following terms.

carbohydrate

chiral molecule

Haworth structure

reducing sugar

polysaccharide

LEARNING EXERCISES

CLASSIFICATION OF CARBOHYDRATES (14.1)

A. Identify the following monosaccharides as aldo-or ketotrioses, tetroses, pentoses, or hexoses:

1.
```
CH2OH
 |
C=O
 |
CH2OH
```

2.
```
HC=O
 |
HCOH
 |
HCOH
 |
HOCH
 |
CH2OH
```

3.
```
CH2OH
 |
C=O
 |
HOCH
 |
HOCH
 |
HCOH
 |
CH2OH
```

4.
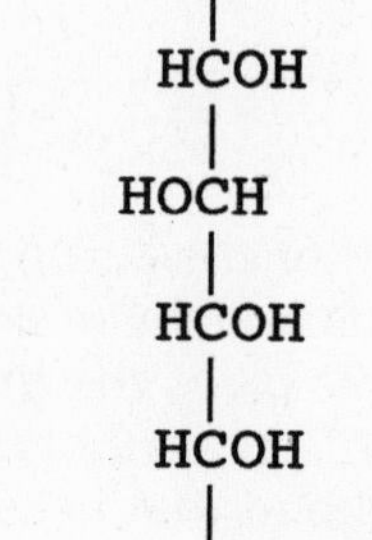

5.
```
HC=O
 |
HCOH
 |
HCOH
 |
CH2OH
```

1. ____________________
2. ____________________
3. ____________________
4. ____________________
5. ____________________

Answers: A. 1. ketotriose 2. aldopentose 3. ketohexose 4. aldohexose 5. aldotetrose

B. Complete and balance the equations for the photosynthesis of

1. glucose: ________ + ________ $\longrightarrow$ $C_6H_{12}O_6$ + ________

2. ribose: ________ + ________ $\longrightarrow$ $C_5H_{10}O_5$ + ________

Answers: (1) $6CO_2 + 6H_2O \longrightarrow C_6H_{12}O_6 + 6O_2$
(2) $5CO_2 + 5H_2O \longrightarrow C_5H_{10}O_5 + 5O_2$

STEREOISOMERS (14.2)

Review: A chiral molecule contains one or more carbon atoms that are attached to four different atoms or groups of atoms. Then the molecule is optically active and has a nonsuperimposable mirror image.

C. State whether each of the following molecules is chiral or achiral:

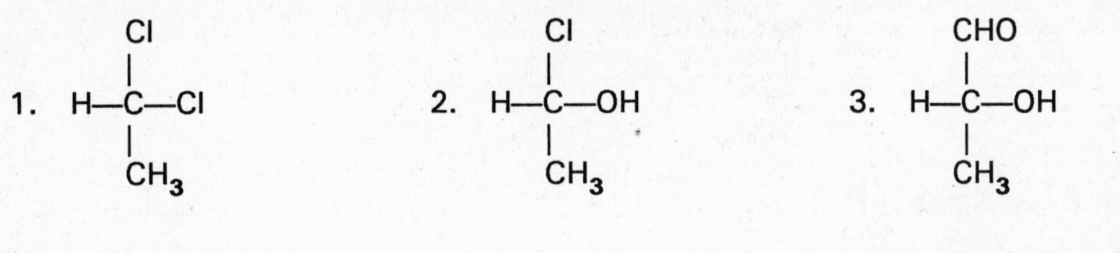

Answers: 1. achiral 2. chiral 3. chiral

D. Identify the following as a feature of a (A) chiral or (B) achiral compound.

1. symmetrical molecule ____________ 2. optically active____________

3. contains a chiral carbon____________ 4. identical mirror images____________

Answers: 1. B 2. A 3. A 4. B

CHIRAL SUGARS (14.3)

Review: In the Fischer projection of a monosaccharide, the chiral -OH farthest from the carbonyl group (C═O) is on the left side in the L-isomer, and on the right side in the D-isomer.

E. Identify the optical isomer of the following sugars by placing a D- or L- in front of the name:

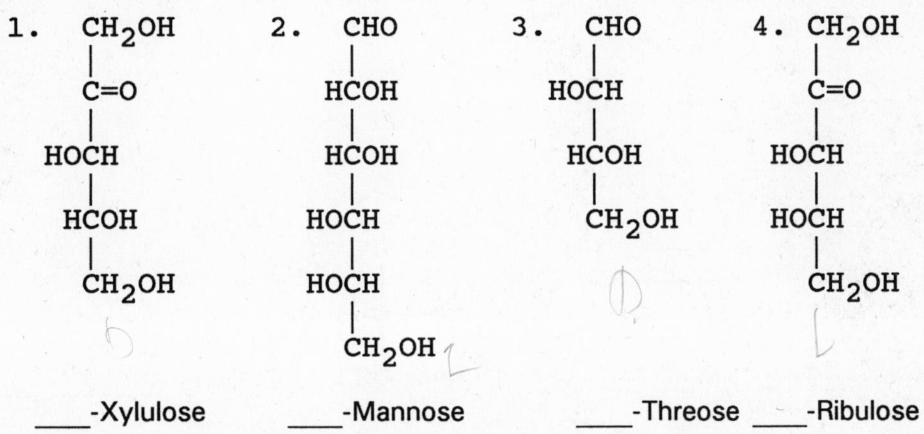

____-Xylulose ____-Mannose ____-Threose ____-Ribulose

Answers:. 1. D-Xylulose 2. L-Mannose 3. D-Threose 4. L-Ribulose

F. Write the mirror image of each of the sugars in part E and gives its D- or L-name.

1. 2. 3. 4.

____-Xylulose ____-Mannose ____-Threose ____-Ribulose

Answers:

1. CH_2OH | C=O | HCOH | HOCH | CH_2OH

L-Xylulose

2. CHO | HOCH | HOCH | HCOH | HCOH | CH_2OH

D-Mannose

3. CHO | HCOH | HOCH | CH_2OH

L-Threose

4. CH_2OH | C=O | HCOH | HCOH | CH_2OH

D-Ribulose

MONOSACCHARIDES AND THEIR HAWORTH STRUCTURES (14.4)

Review: In solution, the hexoses, six-carbon sugars, and pentoses, five-carbon sugars, exist primarily as rings rather than open chains. The rings are formed when an -OH group (usually on carbon 5 in glucose) adds to the carbonyl (C=O) to give a hemiacetal group.

1 HC=O
2 HCOH
3 HOCH
4 HCOH
5 HCOH
6 CH_2OH

D-Glucose (open chain)

⇌

6 CH_2OH, 5 C—O (hemiacetal group), 4 C (H, OH), 3 C (OH, H), 2 C (H, OH), 1 C (H, OH)

hemiacetal group

α-D-Glucose (cyclic)

G. In the structures for glucose, describe the following:

1. The functional groups present in the open chain.

2. The functional groups present in the cyclic form.

3. The reason why glucose is a reducing sugar.

Answers: 1. one aldehyde group, several hydroxyl (OH) groups
2. one hemiacetal group(ether link and alcohol on same carbon) and several hydroxyl groups
3. For the short time that the open chain is present during mutarotation, an aldehyde group becomes available for oxidation.

H. Write the Haworth structures (*α* and ß anomers, if any) for the following:

1. Glucose 2. Galactose 3. Fructose

Answers: 1. α-Glucose

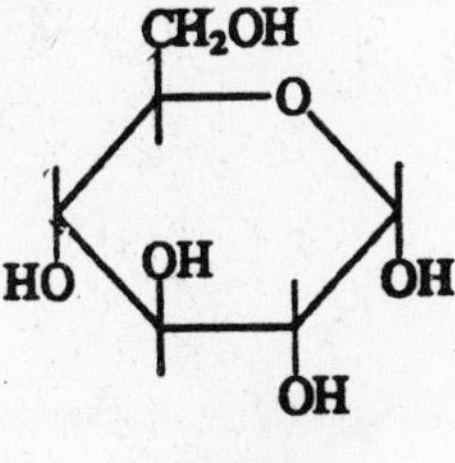

2. α-Galactose

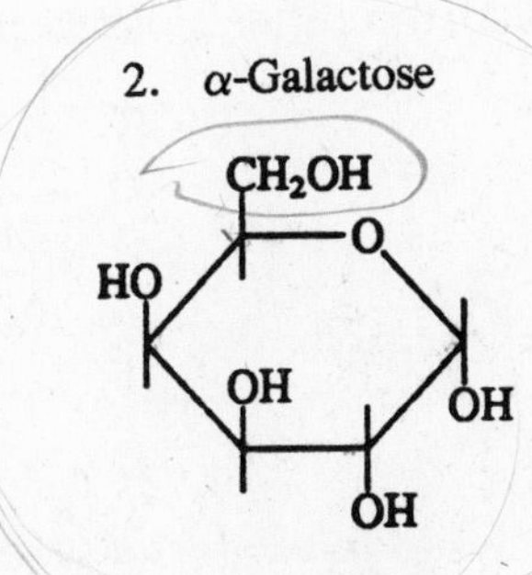

3. α-Fructose

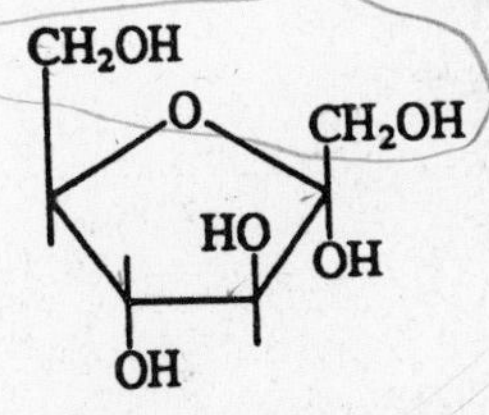

ß-Glucose

ß-Galactose

ß-Fructose

PROPERTIES OF MONOSACCHARIDES (14.5)

I. *Essay:*

1. What changes occur when a reducing sugar reacts with Benedict's reagent?

2. What is a glycosidic bond?

Essay Answers: 1. The carbonyl group of the reducing sugar is oxidized to a carboxylic acid group while the Cu^{2+} ion in Benedict's reagent is reduced to Cu^{+} which forms a brick-red solid.
2. A glycosidic bond forms between the OH of the hemiacetal group of a sugar with the alcohol (-OH) of another compound, usually another sugar.

DISACCHARIDES (14.6)

Review: A disaccharide forms when two monosaccharides link together through a glycosidic bond.

$$\text{monosaccharide}_1 + \text{monosaccharide}_2 \rightleftharpoons \text{disaccharide} + H_2O$$

In the disaccharide maltose, two glucose units are linked by an α-1,4-glycosidic bond. The α-1,4 indicates that the OH of the alpha anomer at carbon 1 was bonded to the OH on carbon 4 of the other glucose molecule.

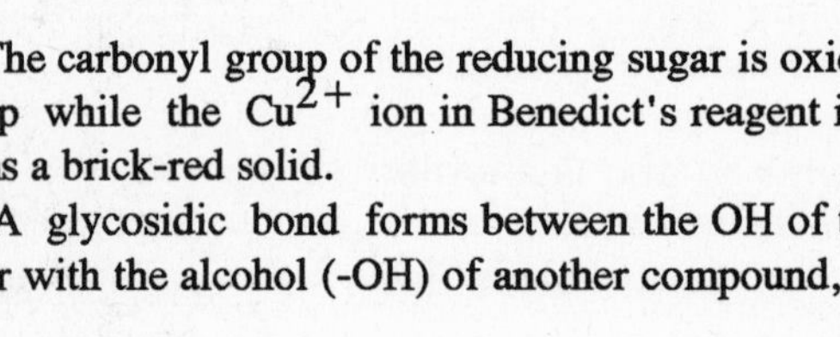

CH₂OH O α CH₂OH O OH 1 4 HO OH OH HO OH OH OH

When a disaccharide is hydrolyzed by water, a glucose unit and one other monosaccharide result.

Maltose + H_2O ⟶ Glucose + Glucose
Lactose + H_2O ⟶ Glucose + Galactose
Sucrose + H_2O ⟶ Glucose + Fructose

J. List the monosaccharides found in the following sugars:

1. Lactose ____________ ____________

2. Maltose ____________ ____________

3. Sucrose ____________ ____________

Answers: 1. glucose + galactose 2. glucose + glucose 3. glucose + fructose

K. For the following disaccharides, state (a) the monosaccharide units, (b) the type of glycosidic bond, (c) the name of the disaccharide, and (d) if it is a reducing sugar.

1.

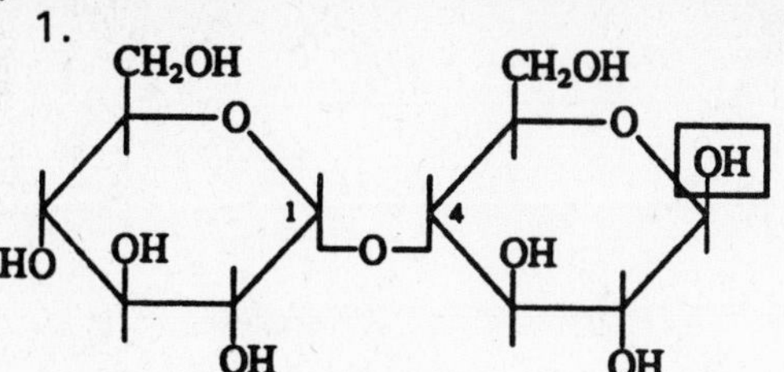

a.____________

b.____________

c.____________

d.____________

2.

CH₂OH O HO OH OH 1 O 4 CH₂OH O OH OH OH

a.____________

b.____________

c.____________

d.____________

3.

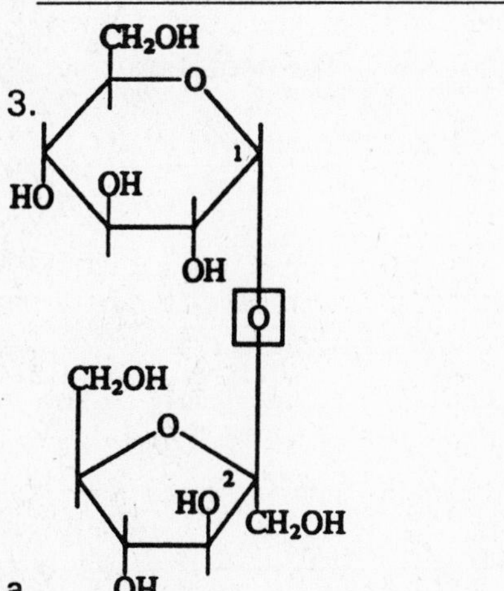

a.____________

b.____________

c.____________

d.____________

4.

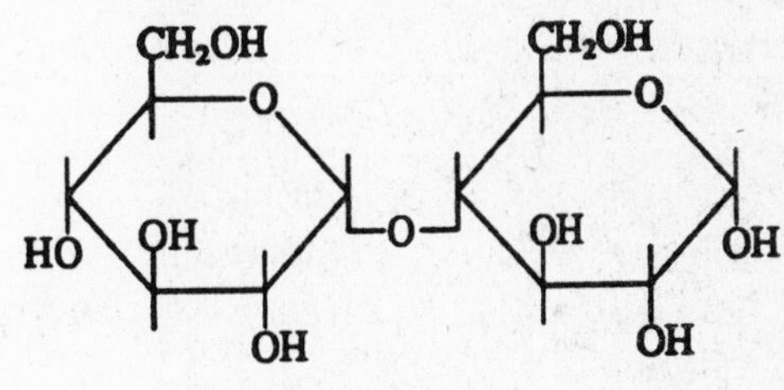

a.____________

b.____________

c.____________

d.____________

Answers: 1.(a) two glucose units, (b) α-1,4 glycosidic bond, c. ß-maltose, (d) reducing sugar
2.(a) galactose + glucose, (b) ß-1,4 glycosidic bond, (c) α-lactose, (d) reducing sugar
3.(a) fructose + glucose, (b) α-1,2 glycosidic bond, (c) sucrose, (d) not a reducing sugar
4.(a) two glucose units, (b) α-1,4 glycosidic bond, (c) α-maltose, (d) reducing sugar

POLYSACCHARIDES 14.7)

Review: The polysaccharides amylose and amylopectin (starch from plants), glycogen (animal starch), and cellulose (plant structural material-fiber) are all polymers of glucose units. Except for cellulose, the glucose units are linked by α-1,4-glycosidic bonds. Amylose is a straight-chain polymer, while amylopectin and glycogen have branches with α-1,6-glycosidic bonds.

L. List the monosaccharides and describe the glycosidic bonds in each of the following carbohydrates:

	Monosaccharides	Type(s) of glycosidic bonds
1. amylose	________	________
2. amylopectin	________	________
3. glycogen	________	________
4. cellulose	________	________

Answers: 1. glucose; α-1,4-glycosidic bonds 2. glucose; α-1,4- and α-1,6-glycosidic bonds 3. glucose; α-1,4- and α-1,6-glycosidic bonds 4. glucose; ß-1,4-glycosidic bonds

TESTS FOR CARBOHYDRATES 14.8)

M. Indicate the results of the following tests as (+) positive (-) negative

	Iodine	Benedict's	Fermentation
1. starch	________	________	________
2. glucose	________	________	________
3. lactose	________	________	________
4. maltose	________	________	________
5. galactose	________	________	________
6. glycogen	________	________	________
7. sucrose	________	________	________

Answers: 1. starch +, −, − 2. glucose −, +, + 3. lactose −, +, − 4. maltose −, +, + 5. galactose −, +, − 6. glycogen +, −, − 7. sucrose −, −, +

SELF-EVALUATION TEST

CARBOHYDRATES

1. The requirements for photosynthesis are
   A. sun
   B. sun and water
   C. water and carbon dioxide
   D. sun, water and carbon dioxide
   E. carbon dioxide and sun

2. What are the products of photosynthesis?
   A. carbohydrates
   B. carbohydrates and oxygen
   C. carbon dioxide and oxygen
   D. carbohydrates and carbon dioxide
   E. water and oxygen

3. The name "carbohydrate" came from the fact that
   A. carbohydrates are hydrates of water.
   B. carbohydrates contain hydrogen and oxygen in a 2:1 ratio.
   C. carbohydrates contain a great quantity of water.
   D. all plants produce carbohydrates.
   E. carbon and hydrogen atoms are abundant in carbohydrates.

4. What functional group are in the open chains of monosaccharides?
   A. hydroxyl groups
   B. aldehyde groups
   C. ketone groups
   D. hydroxyl and aldehyde or ketone groups
   E. hydroxyl groups and a hemiacetal group

5. What is the classification of the following sugar?

$$\begin{array}{c} CH_2OH \\ | \\ C=O \\ | \\ HCOH \\ | \\ CH_2OH \end{array}$$

   A. aldotriose
   B. ketotriose
   C. aldotetrose
   D. ketotetrose
   E. ketopentose

Questions 6,7,8, 9 and 10 refer to

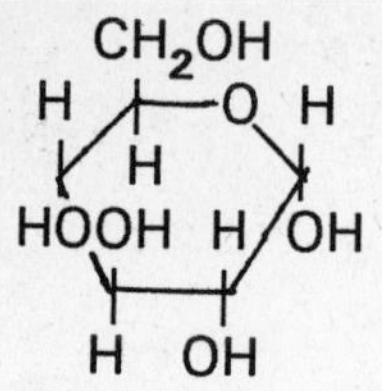

```
     CH2OH
  H ⁄|—O  H
   ⁄ H    \
 HOOH H/OH
   \|——|/
   H  OH
```

6. It is the ring form of an
   A. aldotriose
   B. ketopentose
   C. aldopentose
   D. aldohexose
   E. aldoheptose

7. This is the Haworth structure of
   A. fructose B. glucose C. ribose D. glyceraldehyde E. galactose

8. It is at least one of the products of the complete hydrolysis of
   A. maltose B. sucrose C. lactose D. glycogen E. All of these

9. A Benedict's test with this sugar would
   A. be positive.
   B. be negative.
   C. produce a blue precipitate.
   D. produce no color change.
   E. produce a silver mirror.

10. It is the monosaccharide unit used to build polymers of
    A. amylose.
    B. amylopectin.
    C. cellulose.
    D. glycogen.
    E. All of these

Identify the carbohydrate described in 11-15 as one of the following:

A. maltose B. sucrose C. cellulose D. amylopectin E. glycogen

11.____A disaccharide that is not a reducing sugar.

12.____A disaccharide that occurs as a breakdown product of amylose.

13.____A carbohydrate that is produced as a storage form of energy in plants.

14.____The storage form of energy in humans.

15.____ A carbohydrate that is used for structural purposes by plants.

For questions 16-20 select answers from

A. amylose B. cellulose C. glycogen D. lactose E. sucrose

16.____A polysaccharide composed of many glucose units linked by α-1,4-glycosidic bonds.

17.____A sugar composed of glucose and galactose.

18.____A sugar composed of both α-1,4- and α-1,6-glycosidic bonds.

19.____A sugar that has no anomeric forms.

20.____A carbohydrate composed of ß-1,4-glycosidic bonds.

21.____A sugar composed of glucose and fructose.

For questions 21-25, select answers from the following:

A. Amylose B. Lactose C. Sucrose D. Maltose

22. _____Gives a positive Benedict's test, but is negative for fermentation.

23. _____Gives a negative Benedict's test and a positive fermentation test.

24. _____Gives a positive iodine test.

25. _____Gives a positive Benedict's test and undergoes fermentation.

ANSWERS TO THE SELF-EVALUATION TEST

1. D	6. D	11. B	16. A	21. E
2. B	7. B	12. A	17. D	22. B
3. B	8. E	13. D	18. C	23. C
4. D	9. A	14. E	19. E	24. A
5. D	10. E	15. C	20. B	25. D

SCORING THE SELF-EVALUATION TEST

25 questions 4 points each 100 points

CHAPTER 15
LIPIDS

KEY CONCEPTS

1. Lipids are a family of nonpolar compounds that are not very soluble in water. Several kinds of lipids contain fatty acids, long-chain carboxylic acids, that may be saturated or unsaturated. Unsaturated fatty acids have lower melting points than saturated fatty acids.

2. Waxes and triglycerides consist of fatty acids bonded through ester bonds to alcohols. A long-chain alcohol is used in a wax; triglycerides use glycerol.

3. When hydrogen is added to unsaturated fatty acids, they are converted to saturated fats with higher melting points. Triglycerides undergo hydrolysis in acid or by a digestive enzyme to produce fatty acids and glycerol. In saponification, a fat reacts with a strong base to produce the salts of the fatty acids (soaps) and glycerol.

4. Phospholipids are similar to triglycerides in which one of the fatty acids is replaced by a phosphate attached to an amino alcohol. This change in structure produces a polar portion within the phospholipid, an important feature in their role in cellular membranes. Related compounds includes sphingosines and glycolipids.

5. The steroids and terpenes are types of lipids that contain a steroid portion or isoprene units rather than fatty acids. They play important roles in the production of vitamins and hormones.

KEY WORDS

Use complete sentences to define the following terms:

lipid

fatty acid

triglyceride

saponification

phospholipid

steroid

LEARNING EXERCISES

FATTY ACIDS (15.1)

A. Draw the structural formulas of linoleic acid, stearic acid, and oleic acid.

linoleic acid

stearic acid

oleic acid

Which of these three fatty acids

1. is the most saturated ________________
2. is the most unsaturated ________________
3. has the lowest melting point ________________
4. has the highest melting point ________________
5. is found in vegetables ________________
6. is from animal sources ________________

Answers: linoleic acid $CH_3(CH_2)_4CH{=}CHCH_2CH{=}CH(CH_2)_7\overset{O}{\overset{\|}{C}}OH$

stearic acid $CH_3(CH_2)_{16}\overset{O}{\overset{\|}{C}}OH$

oleic acid $CH_3(CH_2)_7CH{=}CH(CH_2)_7\overset{O}{\overset{\|}{C}}OH$

1. stearic acid
2. linoleic acid
3. linoleic acid
4. stearic acid
5. linoleic and oleic acid
6. stearic acid

WAXES, FATS, AND OILS (15.2)

B. Write the formula of the wax formed by the reaction of palmitic acid, $CH_3(CH_2)_{14}COOH$ and cetyl alcohol, $CH_3(CH_2)_{14}CH_2OH$.

Answer: $CH_3(CH_2)_{14}\overset{\overset{\displaystyle O}{\|}}{C}OCH_2(CH_2)_{14}CH_3$

C. Write the the structure and name of the triglyceride formed from the following:

1. glycerol and three palmitic acids, $CH_3(CH_2)_{14}COOH$

2. glycerol and three myristic acids, $CH_3(CH_2)_{12}COOH$

Answers:

1. 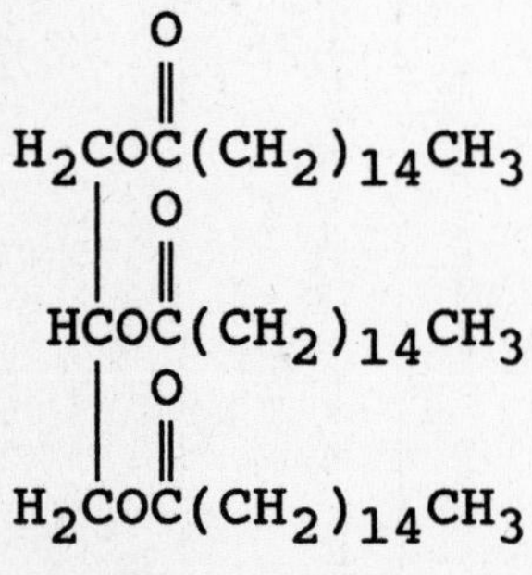
$$\begin{array}{l} H_2CO\overset{\overset{\displaystyle O}{\|}}{C}(CH_2)_{14}CH_3 \\ \ \ | \\ HCO\overset{\overset{\displaystyle O}{\|}}{C}(CH_2)_{14}CH_3 \\ \ \ | \\ H_2CO\overset{\overset{\displaystyle O}{\|}}{C}(CH_2)_{14}CH_3 \end{array}$$

Tripalmitin

2. 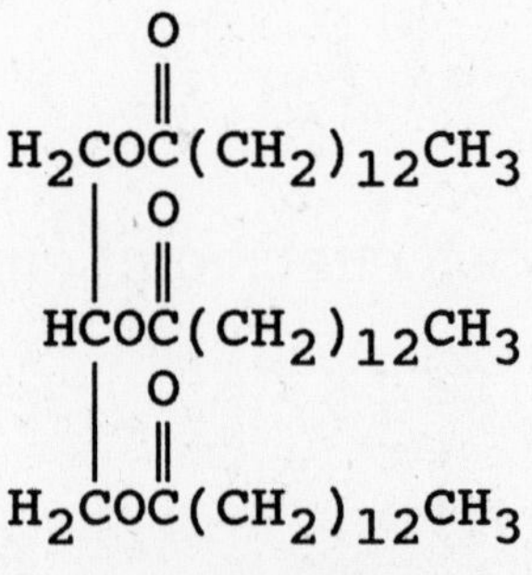
$$\begin{array}{l} H_2CO\overset{\overset{\displaystyle O}{\|}}{C}(CH_2)_{12}CH_3 \\ \ \ | \\ HCO\overset{\overset{\displaystyle O}{\|}}{C}(CH_2)_{12}CH_3 \\ \ \ | \\ H_2CO\overset{\overset{\displaystyle O}{\|}}{C}(CH_2)_{12}CH_3 \end{array}$$

Trimyristin

D. Write the structural formulas of the following triglycerides:

1. Tristearin

2. Triolein

Answers:

1. 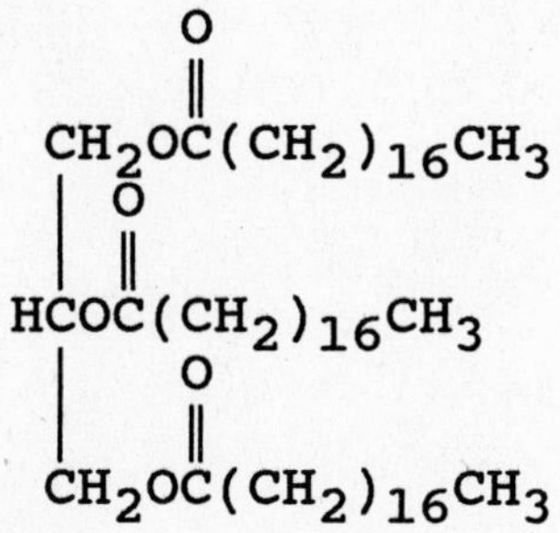
$$\begin{array}{l} CH_2OC(=O)(CH_2)_{16}CH_3 \\ | \\ HCOC(=O)(CH_2)_{16}CH_3 \\ | \\ CH_2OC(=O)(CH_2)_{16}CH_3 \end{array}$$

2.
$$\begin{array}{l} CH_2OC(=O)(CH_2)_7CH{=}CH(CH_2)_7CH_3 \\ | \\ HC{-}OC(=O)(CH_2)_7CH{=}CH(CH_2)_7CH_3 \\ | \\ CH_2OC(=O)(CH_2)_7CH{=}CH(CH_2)_7CH_3 \end{array}$$

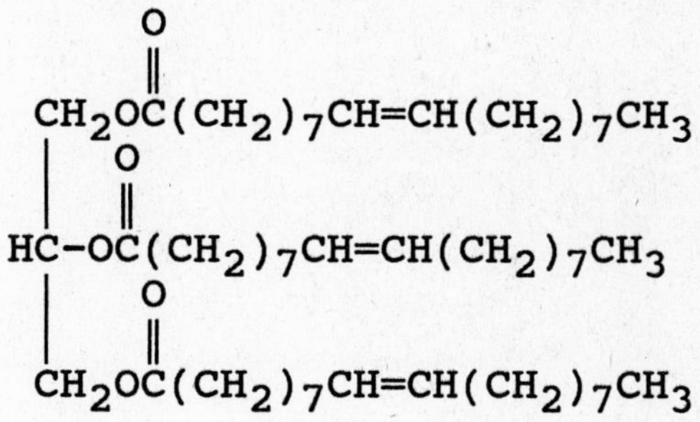

Properties of triglycerides (15.3)

E. Write the equations for the following reactions of triolein:

1. hydrogenation

2. acid hydrolysis with HCl

3. base hydrolysis with NaOH

Answers:

1.

$$\begin{array}{l} CH_2OC(=O)(CH_2)_7CH{=}CH(CH_2)_7CH_3 \\ CH{-}OC(=O)((CH_2)_7CH{=}CH(CH_2)_7CH_3 \\ CH_2OC(=O)(CH_2)_7CH{=}CH(CH_2)_7CH_3 \end{array} + 3H_2 \xrightarrow{Ni} \begin{array}{l} CH_2OC(=O)(CH_2)_{16}CH_3 \\ CHOC(=O)(CH_2)_{16}CH_3 \\ CH_2OC(=O)(CH_2)_{16}CH_3 \end{array}$$

2.

$$\begin{array}{l} CH_2OC(=O)(CH_2)_7CH{=}CH(CH_2)_7CH_3 \\ CH{-}OC(=O)((CH_2)_7CH{=}CH(CH_2)_7CH_3 \\ CH_2OC(=O)(CH_2)_7CH{=}CH(CH_2)_7CH_3 \end{array} + 3H_2O \xrightarrow{H^+} \begin{array}{l} CH_2OH \\ CHOH \\ CH_2OH \end{array} + 3\ HOC(=O)(CH_2)_7CH{=}CH(CH_2)_7CH_3$$

3.

$$\begin{array}{l} CH_2OC(=O)(CH_2)_7CH{=}CH(CH_2)_7CH_3 \\ CH{-}OC(=O)((CH_2)_7CH{=}CH(CH_2)_7CH_3 \\ CH_2OC(=O)(CH_2)_7CH{=}CH(CH_2)_7CH_3 \end{array} + 3NaOH \longrightarrow \begin{array}{l} CH_2OH \\ CHOH \\ CH_2OH \end{array} + 3\ Na^{+}\,{}^{-}OC(=O)(CH_2)_7CH{=}CH(CH_2)_7CH_3$$

PHOSPHOLIPIDS (15.4)

F. Consider the following phosphoglyceride.

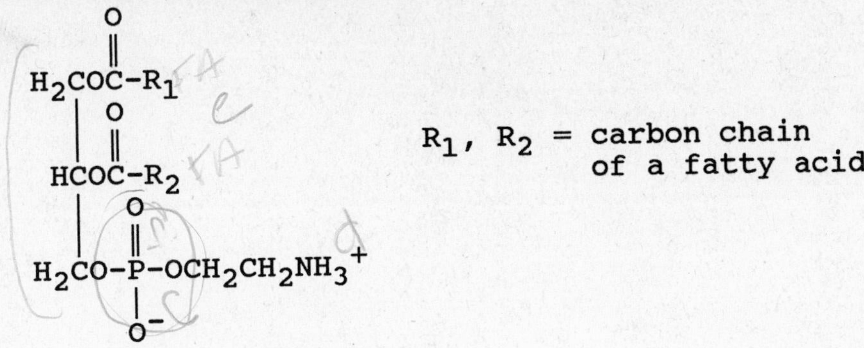

1. On the above structure, indicate the
   a. two fatty acids.
   b. the part from the glycerol molecule.
   c. the phosphate section.
   d. the amino alcohol group.
   e. the nonpolar region.
   f. the polar region.

2. What is the name of the amino alcohol group? ______________

3. What is the name of the phosphoglyceride? ______________

4. Why is a phosphoglyceride more soluble in water than most lipids?

Answers:

glycerol: $H_2COC(=O)-R_1$, $HCOC(=O)-R_2$, $H_2CO-P(=O)(O^-)-OCH_2CH_2NH_3^+$

- R_1, R_2: fatty acids (nonpolar region)
- $-P(=O)(O^-)-$: phosphate
- $OCH_2CH_2NH_3^+$: amino alcohol (polar region)

2. ethanolamine 3. ethanolamine phosphoglyceride 4. The polar portion of the phosphoglyceride is attracted to water which makes this type of lipid more soluble in water compared to other lipids.

G. Draw the structure of a phosphoglyceride that is formed from two molecules of palmitic acid and serine, an amino alcohol:

$CH_3(CH_2)_{14}COOH$
palmitic acid

$HO\text{-}CH_2\text{—}CH(NH_3^+)\text{—}COO^-$
serine

Answer:

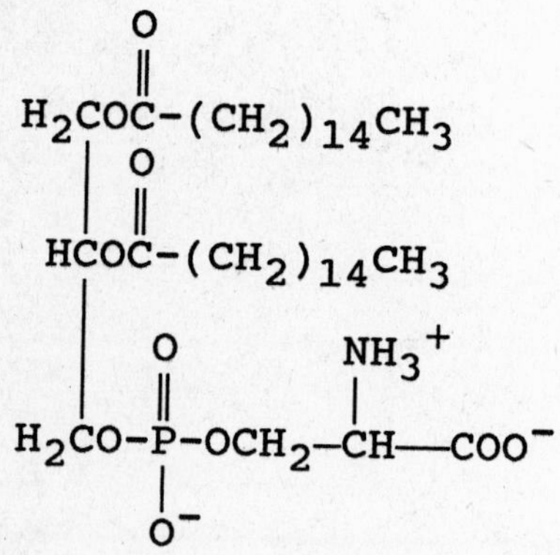

$$
\begin{array}{l}
H_2C\text{—}O\text{—}C(=O)\text{—}(CH_2)_{14}CH_3 \\
\;| \\
HC\text{—}O\text{—}C(=O)\text{—}(CH_2)_{14}CH_3 \\
\;| \\
H_2C\text{—}O\text{—}P(=O)(O^-)\text{—}OCH_2\text{—}CH(NH_3^+)\text{—}COO^-
\end{array}
$$

TERPENES AND STEROIDS (15.5)

H. Write the structure for the steroid nucleus:

Answer:

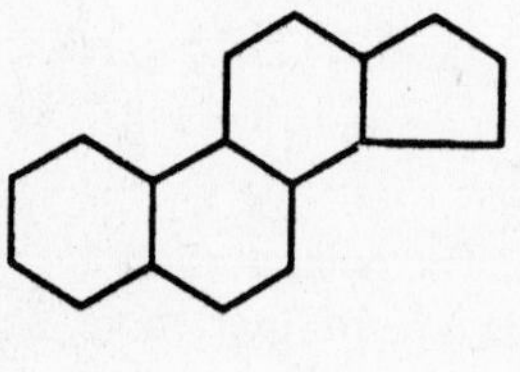

I. Match one of the lipid classes in the second column with the lipid described in the first column

	Lipid	Class of lipid
1. ____	tristearin	A. wax
2. ____	vitamin A	B. triglyceride
3. ____	a compound containing a long chain alcohol and a fatty acid	C. phospholipid
4. ____	cholesterol	D. glycolipid
5. ____	cortisone	E. sphingolipid
6. ____	vegetable oil	F. steroid
7. ____	a compound of glycerol, two fatty acids, phosphate and choline	G. terpene
8. ____	vitamin D	
9. ____	a compound of sphingosine, fatty acid, phosphate and choline	
10. ____	a compound of sphingosine, fatty acid, and galactose	

Answers: 1. B 2. G 3. A 4. F 5. F 6. B 7. C 8. F 9. E 10. D

SELF-EVALUATION TEST

LIPIDS

1. An ester of a fatty acid is called a
   A. carbohydrate B. lipid C. protein D. oxyacid E. soap

2. A fatty acid that is unsaturated is usually
   A. from animal sources and liquid at room temperature.
   B. from animal sources and solid at room temperature.
   C. from vegetable sources and liquid at room temperature.
   D. from vegetable sources and solid at room temperature.
   E. from both vegetable and animal sources and solid at room temperature.

3. The structural formula $CH_3(CH_2)_{14}\overset{\overset{O}{\|}}{C}OH$ is a

   A. unsaturated fatty acid B. saturated fatty acid C. wax
   D. triglyceride E. terpene

For questions 4,5, 6 and 7, consider the following compound:

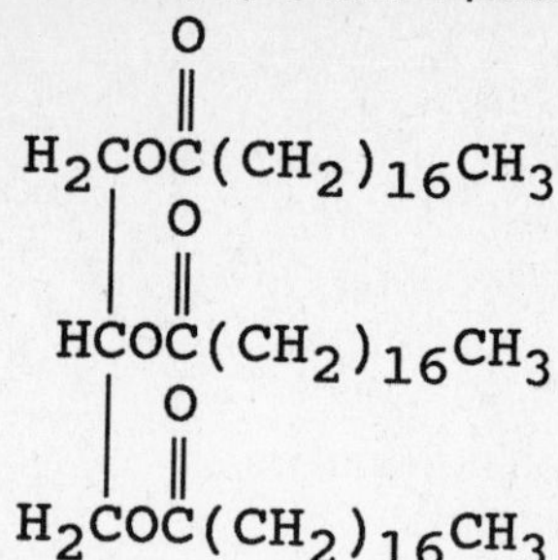

4. This compound belongs in the family called
A. wax
B. triglyceride
C. phosphoglyceride
D. sphingolipid
E. steroid

5. The molecule shown above was formed by
A. esterification
B. hydrolysis (acid)
C. saponification
D. emulsification
E. oxidation

6. If this molecule were reacted with strong base such as NaOH, the products would be
A. glycerol and fatty acids.
B. glycerol and water.
C. glycerol and soap.
D. an ester and salts of fatty acids.
E. an ester and fatty acids.

7. The compound would be expected to be
A. saturated, and a solid at room temperature.
B. saturated, and a liquid at room temperature.
C. unsaturated, and a solid at room temperature.
D. unsaturated, and a liquid at room temperature.
E. supersaturated, and a liquid at room temperature.

8. Which are found in phospholipids?
A. fatty acids
B. glycerol
C. a nitrogen compound
D. phosphate
E. All of these

For questions 9 and 10, consider the following reaction:

Triglyceride + 3NaOH ⟶ 3 sodium salts of fatty acids and glycerol

9. The reaction of a triglyceride with a strong base such as NaOH is called
A. esterification
B. lipogenesis
C. hydrolysis
D. saponification
E. ß-oxidation

10. What is another name for the sodium salts of the fatty acids?
A. margarine
B. fat substitutes
C. soaps
D. perfumes
E. vitamins

For questions 11, 12, 13, 14, 15, and 16, consider the following phosphoglyceride.

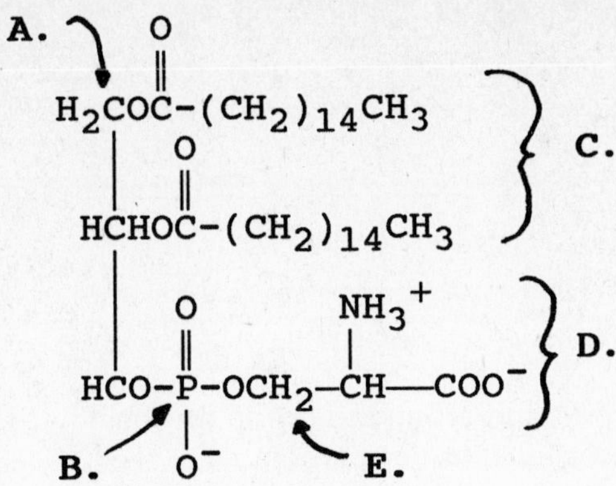

11.____The glycerol portion.

12.____The phosphate portion.

13.____The amino alcohol.

14.____The polar region.

15.____The nonpolar region.

16. What type of phosphoglyceride is this structure?
   A. choline   B. cephalin   C. sphingomyelin   D. glycolipid   E.cerebroside

Classify the following lipids as

A. wax   B. triglyceride   C. phosphoglyceride   D. steroid   E. terpene

17. ____ cholesterol

18. ____ vitamin A

19. ____ $CH_3(CH_2)_{14}\overset{\overset{O}{\|}}{C}O(CH_2)_{30}CH_3$

20. ____ an ester of glycerol with three palmitic acids

ANSWERS TO THE SELF-EVALUATION TEST

1. B	6. C	11. A	16. B
2. C	7. A	12. B	17. D
3. B	8. E	13. E	18. E
4. B	9. D	14. D	19. A
5. A	10. C	15. C	20. B

SCORING THE SELF-EVALUATION TEST

20 questions   5 points each   100 points

CHAPTER 16

PROTEINS

KEY CONCEPTS

1. Amino acids are compounds with a carboxylic group and an amino group on the alpha carbon. They differ only by a side group on the alpha (α)-carbon.

2. A complete protein in the diet or a group of complementary proteins provides all the essential amino acids, amino acids not synthesized by the body.

3. Amino acids exist in solution as dipolar ions called zwitterions. At the isoelectric point, the charge balance is zero, whereas in solutions with pH below or above the isoelectric point, amino acids carry a net positive or negative charge.

4. A peptide consists of amino acids that form a peptide (amide) bond between the carboxylic acid group of one amino acid and the amino group of the next.

5. The primary structure of a protein is the order of the amino acids linked by peptide bonds.

6. The secondary structures of proteins include a coiling of a protein chain as an alpha helix held in place by hydrogen bonding between the coils. Secondary structures are also formed by hydrogen bonding between several protein chains in structures such as the pleated sheet and the triple helix.

7. Tertiary protein structures give a three-dimensional shape produced through cross-links such as salt bridges and disulfide bonds that form between side chains of amino acids in the protein.

8. When proteins are classified by composition, they are grouped as simple when they contain protein only, or conjugated when they contain a nonprotein such as a carbohydrate or metal ion. Another type of classification groups proteins by their biological functions such as structural, catalytic or transport protein.

9. Denaturation, (the destruction of protein structure) occurs when a protein is heated or placed in acid or base, alcohol, or heavy-metal salts. Proteins and amino acids may be detected through the biuret test, the ninhydrin test, the xanthoproteic test and the sulfur test.

KEY WORDS

Using complete sentences, define the following terms:

amino acid

zwitterion

peptide bond

protein

secondary structure

tertiary structure

denaturation

LEARNING EXERCISES

AMINO ACIDS (16.1)

Review: An amino acid consists of a carboxylic acid group, an amino group, and a side chain (R group) on the alpha carbon.

side group

$$\begin{array}{ccccc} & & R & & O \\ & & | & & \| \\ H_2N & — & C & — & COH \\ & & | & & \\ & & H & & \end{array}$$

amino group — **carboxylic acid group** — **alpha carbon**

Nonpolar amino acids contain hydrocarbon side chains, while polar amino acids contain electronegative atoms such as O (—OH) or S (—SH). Acidic side chains contain a carboxylic acid group (—COOH), and basic side chains contain an amino group ($—NH_2$).

A. Using the appropriate R group, complete the structural formula of each of the following amino acids. Indicate whether the amino acid would be polar, nonpolar, acidic, or basic.

glycine (R = –H)

$$H_2N-\overset{|}{\underset{\underset{\displaystyle H}{|}}{C}}-\overset{\overset{\displaystyle O}{\|}}{C}OH$$

____

alanine (R = $-CH_3$)

$$H_2N-\overset{|}{\underset{\underset{\displaystyle H}{|}}{C}}-\overset{\overset{\displaystyle O}{\|}}{C}OH$$

____

serine (R = $-CH_2OH$)

$$H_2N-\overset{|}{\underset{\underset{\displaystyle H}{|}}{C}}-\overset{\overset{\displaystyle O}{\|}}{C}OH$$

____

aspartic acid (R = $-CH_2\overset{\overset{O}{\|}}{C}OH$)

$$H_2N-\overset{|}{\underset{\underset{\displaystyle H}{|}}{C}}-\overset{\overset{\displaystyle O}{\|}}{C}OH$$

____

Answers:

Glycine, $H_2N-\overset{\overset{\displaystyle H}{|}}{\underset{\underset{\displaystyle H}{|}}{C}}-\overset{\overset{\displaystyle O}{\|}}{C}OH$ nonpolar; alanine, $H_2N-\overset{\overset{\displaystyle CH_3}{|}}{\underset{\underset{\displaystyle H}{|}}{C}}-\overset{\overset{\displaystyle O}{\|}}{C}OH$ nonpolar

serine, $H_2N-\overset{\overset{\overset{\displaystyle OH}{|}}{\overset{\displaystyle CH_2}{|}}}{\underset{\underset{\displaystyle H}{|}}{C}}-\overset{\overset{\displaystyle O}{\|}}{C}OH$ polar; aspartic acid, $H_2N-\overset{\overset{\overset{\displaystyle COOH}{|}}{\overset{\displaystyle CH_2}{|}}}{\underset{\underset{\displaystyle H}{|}}{C}}-\overset{\overset{\displaystyle O}{\|}}{C}OH$ acidic

ESSENTIAL AMINO ACIDS (16.2)

B. Seeds, vegetables, and legumes are typically low in one or more of the essential amino acids, tryptophan, isoleucine and lysine.

	tryptophan	isoleucine	lysine
sesame seeds	OK	LOW	LOW
sunflower seeds	OK	OK	LOW
garbanzo beans	LOW	OK	OK
rice	OK	OK	LOW
cornmeal	OK	OK	LOW

Indicate whether the following proteins are complimentary or not?

1. _____sesame seeds and sunflower seeds
2. _____sunflower seeds and garbanzo beans
3. _____sunflower seeds, sesame seeds and garbanzo beans
4. _____sesame seeds and garbanzo beans
5. _____sunflower seeds and milk
6. _____rice and garbanzo beans
7. _____rice and cornmeal

Answers: 1. not complementary; both are low in lysine
2. complementary 3. complementary 4. complementary
5. complementary 6. complementary 7. not complementary; both low in lysine

PROPERTIES OF AMINO ACIDS (16.3)

Review: Amino acids exist in a salt-like form called a *zwitterion*. At its isoelectric point, the charge balance is zero. However, when placed in solutions having a pH below or above the pH of the isoelectric point, the amino acid has a net overall positive or negative charge.

Example: Glycine has an isoelectric point at a pH of 6.0. In more acidic solutions, it has a net positive charge, and in more basic solutions, a net negative charge.

$$H_3N^+\text{-}CH_2\text{-}COOH \xleftarrow{H^+} H_3N^+\text{-}CH_2\text{-}COO^- \xrightarrow{OH^-} H_2N\text{-}CH_2\text{-}COO^-$$

zwitterion of glycine

C. Write the structure of the amino acids under the given conditions:

	zwitterion	H^+	OH^-
alanine			
serine			

Answers:

alanine	$^{+}NH_3CH(CH_3)COO^-$	$^{+}NH_3CH(CH_3)COOH$	$NH_2CH(CH_3)COO^-$
serine	$^{+}NH_3CH(CH_2OH)COO^-$	$^{+}NH_3CH(CH_2OH)COOH$	$NH_2CH(CH_2OH)COO^-$

PEPTIDES (16.4)

Review: A peptide forms when an amide bond (peptide bond) joins the carboxylic acid group of one amino acid to the amino group of the next. The free amino group is called the *N-terminal* and the free carboxylic acid is called the *C-terminal*.

$$H_2N\text{-}CH_2\text{-}\overset{O}{\overset{\|}{C}}\text{-}\overset{H}{\overset{|}{N}}\text{-}\overset{CH_3}{\overset{|}{C}H}\text{-}COOH$$

N-terminal **peptide bond** **C-terminal**

D. Draw the structural formulas of the following di- and tripeptides.

1. serylglycine

2. cystylvaline

3. Gly—Ser—Cys

Answers:

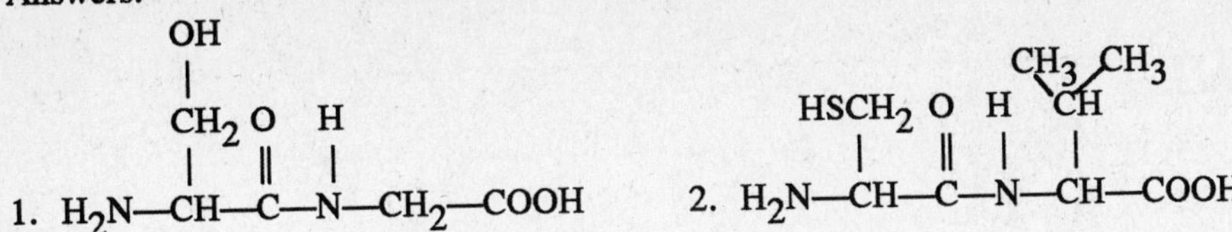

```
                        OH
                        |
          O  H   CH2  O  H   CH2SH
          ‖  |    |   ‖  |    |
3. H2N—CH2—C—N—CH—C—N—CH—COOH
```

PROTEIN STRUCTURE (16.5, 16.6, 16.7)

E. State the level of protein structure represented by each of the following: primary, secondary, tertiary, or quaternary.

1. ________ A disulfide bond joining distant parts of a peptide.
2. ________ Hydrogen bonding to form an alpha (α)-helix.
3. ________ The combination of four protein subunits.
4. ________ Amino acids linked by peptide bonds.
5. ________ Hydrogen bonding to form a pleated-sheet structure.
6. ________ Hydrophilic side groups seeking contact with water.
7. ________ A salt bridge forms between two oppositely charged side chains.
8. ________ Hydrophobic side groups forming a nonpolar center.

Answers: 1. tertiary
2. secondary
3. quaternary
4. primary
5. secondary
6. tertiary
7. tertiary
8. tertiary

CLASSIFICATION OF PROTEINS (16.8)

F. Classify the following as *simple (S)* or *conjugated (C)* proteins.

1.______ antibodies (globulins)

2.______ hemoglobin

3.______ casein, a protein with phosphate

4.______ ribosomes, protein with ribonucleic acid

5.______ tyrosine oxidase, protein with copper

6.______ nail protein

Answers: 1. S 2. C 3. C 4. C 5. C 6. S

G. Match one of the following functions of a protein with the examples below:

structural	contractile	storage	
transport	hormonal	catalytic	antibody

1. __________hemoglobin to carry oxygen in blood
2. __________amylase, an enzyme that hydrolyzes starch
3. __________egg albumin, a protein in egg white
4. __________growth hormone which controls growth
5. __________collagen, a major part of connective tissue
6. __________immunoglobulin
7. __________keratin, a major protein of hair
8. __________lipoprotein, a lipid carrier in blood

Answers 1. transport 2. catalytic 3. storage 4. hormonal
5. structural 6. antibody 7. structural 8. transport

PROPERTIES OF PROTEINS (16.9)

Review: Proteins are denatured (lose their activity) when cross-links of tertiary and sometimes secondary protein structures are destroyed by agents such as heat, acid, base, organic solvents, agitation, heat, and metal ions.

H. Indicate the denaturing agent in the following examples:

A. heat or UV light B. pH change C. organic solvent
D. heavy metal ions E. agitation

1.____ Placing surgical instruments in a 120°C autoclave.

2.____ Whipping cream to make a desert topping.

3.____ Applying tannic acid to a burn.

4.____ Placing $AgNO_3$ drops in the eyes of newborns.

5.____ Using alcohol to disinfect a wound.

6.____ Using lactobacillus bacteria culture to produce acid that converts milk to yogurt.

Answers: 1. A 2. E 3. C 4. D 5. C 6. B

I. Indicate if the following will give positive or negative results in the following tests.

	biuret	ninhydrin	xanthoproteic	sulfur
Ala	____	____	____	____
Cys	____	____	____	____
Ala—Gly	____	____	____	____
Ala—Tyr	____	____	____	____
Ala—Ala—Ala	____	____	____	____
Ala—Try—Gly	____	____	____	____

	biuret	ninhydrin	xanthoproteic	sulfur
Ala	−	+	−	−
Cys	−	+	−	+
Ala-Gly	−	−	−	−
Ala-Tyr	−	−	+	−
Ala-Ala-Ala	+	−	−	−
Ala-Try-Gly	+	−	+	−

SELF-EVALUATION TEST

PROTEINS

1. Which amino acid is nonpolar?
   A. serine B. aspartic acid C. valine D. cysteine E. glutamine

2. Which amino acid will form disulfide cross—links in a tertiary structure?
   A. serine B. aspartic acid C. valine D. cysteine E. glutamine

3. Which amino acid has a basic side chain.
   A. serine B. aspartic acid C. valine D. cysteine E. glutamine

4. All amino acids
   A. have the same side chains.
   B. form zwitterions.
   C. have the same isoelectric points.
   D. show hydrophobic tendencies.
   E. are essential amino acids.

5. Essential amino acids
   A. are the amino acids needed in the diet.
   B. are synthesized by the body.
   C. are found in incomplete proteins.
   D. are the amino acids needed by the body.
   E. are obtained in a vegetarian diet.

Use the following to answer questions for the amino acid alanine in questions 6, 7, 8,and 9.

A. $^{+}H_3N\text{—}\underset{}{\overset{CH_3}{\overset{|}{CH}}}\text{—}COO^{-}$ B. $H_2N\text{—}\overset{CH_3}{\overset{|}{CH}}\text{—}COO\text{–}$ C. $^{+}H_3N\text{—}\overset{CH_3}{\overset{|}{CH}}\text{—}COOH$

6. ____ Alanine in its zwitterion form.

7. ____ Alanine at a low pH.

8. ____ Alanine at a high pH.

9. ____ Alanine at its isoelectric point.

10. The sequence Tyr—Ala—Gly
    A. is a tripeptide.
    B. has two peptide bonds.
    C. has tyrosine as the N—terminal end.
    D. has glycine as the C—terminal end.
    E. All of these.

11. What is the amino acids sequence if the segments Cys-Glu and Phe-Ala-Cys are obtained from the partial hydrolysis of a tetrapeptide?
    A. Phe-Ala-Cys-Cys-Glu
    B. Cys-Glu-Phe-Ala
    C. Cys-Glu-Phe-Ala-Cys
    D. Phe-Ala-Cys-Glu
    E. Cys-Glu-Phe-Ala

12. What type of bond is used to the α-helix structure of a protein?
    A. peptide bond
    B. hydrogen bond
    C. salt bridge
    D. disulfide bond
    E. hydrophobic

13. What type of bonding places portions of the protein chain in the center of a tertiary structure?
    A. peptide bonds
    B. salt bridges
    C. disulfide bonds
    D. hydrophobic attractions
    E. hydrophilic attractions

In questions 14, 15, 16, 17, and 18, identify the protein structural levels that each of the following statements describe.

14. ____ peptide bonds
15. ____ a pleated sheet
16. ____ two or more protein subunits.
17. ____ an α-helix
18. ____ disulfide bonds

A. primary
B. secondary
C. tertiary
D. quaternary
E. pentinary

19. The presence of zinc in the enzyme called alcohol dehydrogenase classifies the protein as
    A. primary B. conjugated C. hormonal D. structural E. secondary

In questions 20, 21, 22, and 24, match the function of a protein with each example.

20. ____ catalytic
21. ____ structural
22. ____ transport
23. ____ storage

A. myoglobin in the muscles
B. α-keratin in skin
C. peptidase for protein hydrolysis
D. casein in milk

24. Denaturation of a protein
    A. occurs at a pH of 7.
    B. causes a change in protein structure.
    C. hydrolyzes a protein.
    D. oxidizes the protein.
    E. adds amino acids to a protein.

25. Which of the following will not cause denaturation?
    A. 0°C
    B. $AgNO_3$
    C. 80°C
    D. ethanol
    E. pH 1

ANSWERS TO THE SELF-EVALUATION TEST

1. C	6. A	11. D	16. D	21. B
2. D	7. C	12. B	17. B	22. A
3. E	8. B	13. D	18. C	23. D
4. B	9. A	14. A	19. B	24. B
5. A	10. E	15. B	20. C	25. A

SCORING THE SELF-EVALUATION TEST

25 questions 4 points each 100 points

CHAPTER 17
ENZYMES, VITAMINS, AND HORMONES

KEY CONCEPTS

1. Enzymes are catalysts that increase the rate of reaction by lowering the energy required for a reaction. Enzymes react with specific substrates because of their unique structural conformation. The lock-and-key theory suggests that there is an active site within an enzyme where the enzyme-catalyzed reaction occurs.

2. A common naming system names enzymes by replacing the ending of the substrate names with the suffix *-ase*. A systematic naming system classifies and names enzymes by six general reaction types. Enzymes consisting of protein are called simple enzymes; enzymes consisting of protein and a cofactor, metal ion or vitamin are called conjugated enzymes.

3. The rate of an enzymatic reaction depends upon the temperature, pH, amount of catalyst, and the amount of substrate. Enzymes are most active at their optimum temperatures and pH.

4. A competitive inhibitor is similar in structure to the substrate and completes for the active site of the enzyme which inhibits the action of that enzyme. A noncompetitive inhibitor changes the structure of the enzyme and prevents the substrate from fitting correctly into the active site.

5. The digestion of carbohydrates, lipids, and proteins is largely a process of hydrolysis carried out by specific enzymes. For carbohydrates such as starch, this action begins in the mouth with the action of salivary amylase; lipids begin their digestion in the small intestine through the action of bile salts and lipases; proteins begin hydrolysis in the stomach through the action of pepsin and later, this is continued in the small intestine by peptidases.

6. Vitamins are organic compounds needed for continued health and must be obtained from the diet. The water-soluble vitamins include vitamins B and C; the fat-soluble vitamins are vitamins A, D, E, and K. Most of the water-soluble vitamins function as coenzymes in metabolic reactions.

7. Hormones which may be peptides, proteins, or steroids stimulate enzyme production in target cells and thereby accelerate metabolic reactions.

8. Hormones are organic compounds necessary for continued health that are secreted by glands in the body and carried by the bloodstream to other target organs where they affect the activity levels of the target cells. Some hormones activate cyclic AMP, increasing the activity level of the cell or increasing the level of enzymatic proteins.

KEY WORDS

Using complete sentences, define the following terms:

enzyme

lock-and-key theory

competitive inhibition

optimum temperature

vitamin

hormone

LEARNING EXERCISES

ENZYME ACTION (17.1)

Review: Enzymes increase the rate of a reaction by lowering the energy required for a substrate to undergo reaction and form product. Binding takes place when the structure of the enzyme complements the structure of the substrate and an enzyme-substrate complex forms. Within the enzyme, there is a region called the active site in which the reaction of the substrate take place. As soon as the product is formed, it is released from the enzyme.

$$E + S \rightleftharpoons ES \longrightarrow E + P$$

Enzyme + Substrate ⇌ Enzyme-Substrate Complex → Enzyme + Product

A. Draw an energy diagram for the decomposition of hydrogen peroxide catalyzed by catalase. Label the energy levels of the reactants and products and the energy required with and without enzyme.

$$2H_2O_2 \longrightarrow O_2 + 2H_2O$$

Energy↑

Reaction progress→

Answer

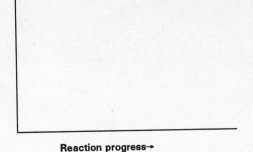

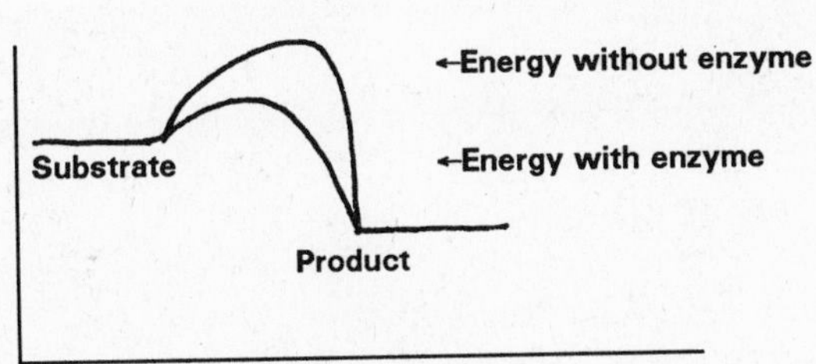

B. Write an equation to illustrate the following:

1. The formation of an enzyme-substrate complex. ______________

2. The conversion of enzyme-substrate complex to product. ______________

Answers: 1. $E + S \rightleftharpoons ES$ 2. $ES \longrightarrow E + P$

NAMES AND CLASSIFICATION OF ENZYMES (17.2)

Review: In the common names of enzymes, the suffix *-ase* replaces the ending of the name of the substrate in the reaction. In the systematic naming system, six reaction types are used to name the enzyme.

C. Match the common names of enzymes with the substrates in the reactions.
A. dehydrogenase B. oxidase C. peptidase D. peroxidase E. esterase

1. ____ Catalyzes the hydrolysis of esters into fatty acids and glycerol.

2. ____ Catalyzes the removal of hydrogen from a substrate.

3. ____ Catalyzes the activation of oxygen so it will combine with a substrate.

4. ____ Catalyzes the decomposition of hydrogen peroxide to water and oxygen.

5. ____ Catalyzes the hydrolysis of peptide bonds in the digestion of proteins.

Answers: 1. E 2. A 3. B 4. D 5. C

D. State the IUPAC classification for enzymes that catalyze the following reactions:
A. oxidoreductase B. transaminase C. hydrolase D. hydrase E. isomerase

1. ____ An enzyme that removes a CO_2 group from a carboxylic acid.

2. ____ An enzyme that transfers an amino (NH_2-) group from an amino acid to an α-keto acid.

3. ____ A carboxylase that uses H_2O to split a disaccharide into two glucose units.

4. ____ An enzyme that converts glucose-1-phosphate into fructose-1-phosphate during glycolysis.

5. ____ A dehydrogenase that causes the oxidation of ethanol to acetaldehyde.

Answers: 1. D 2. B 3. C 4. E 5. A

E. Indicate whether each statement describes a simple or a conjugated enzyme.

1. ____ An enzyme consisting only of protein.

2. ____ An enzyme requiring magnesium ion for activity.

3. ____ An enzyme containing a sugar group.

4. ____ An enzyme that gives only amino acids upon hydrolysis.

5. ____ An enzyme that requires zinc ions for activity.

Answers: 1. simple 2. conjugated 3. conjugated 4. simple 5. conjugated

FACTORS THAT AFFECT ENZYME ACTIVITY (17.3)

F. Match the statements below with the following terms:
A. increase B. decrease C. optimum temperature D. optimum pH E. no change

1. ____ The pH at which an enzyme is most active.

2. ____ The temperature at which an enzyme is most active.

3. ____ Running a metabolic reaction at 85°C.

4. ____ The effect on the enzyme activity when enzyme concentration increases.

5. ____ The effect on enzyme activity at 0°C.

6. ____ Running the reaction at pH 2 when the enzyme's optimum pH is 7.5.

7. ____ Adding more substrate to the reaction flask.

8. ____ Adjusting the pH to optimum pH.

Answers: 1. D 2. C 3. B 4. A 5. B 6. B 7. A 8. A

ENZYME INHIBITION (17.4)

Review: A ***competitive inhibitor*** which has a structure similar to the substrate competes with the substrate for the active site. The inhibition can be reversed by adding more substrate. A ***noncompetitive inhibitor*** alters the shape of the enzyme and interferes with the fit of the substrate at the active site. This type of inhibition cannot be reversed by adding more substrate. ***Irreversible inhibition*** destroys catalytic activity by denaturing the enzyme or forming strong covalent bonds between a reagent and side chains in the active site.

G. Identify the type of inhibition in each of the following as
(A) competitive (B) noncompetitive (C) irreversible

1.____ The inhibitor binds to the surface of the enzyme.

2.____ The inhibitor resembling the substrate molecule blocks the active site on the enzyme.

3.____ The inhibition causes permanent damage to the enzyme with total loss of biological activity.

4.____ The action of this inhibitor can be reversed by adding more substrate.

5.____ Increasing substrate concentration does not change the effect of this inhibition.

6.____ A molecule which will stop bacterial growth and closely resembles the substrate molecule.

Answers: 1. B 2. A 3. C 4. A 5. A

ENZYMES IN DIGESTION (17.5)

H. Complete the table to describe digestive processes for carbohydrates, lipids and proteins.

	Food	*Digestion site(s)*	*Enzyme*	*Products*
1.	amylose	____________	________	________
2.	maltose	____________	________	________
3.	cellulose	____________	________	________
4.	fat	____________	________	________
5.	protein	____________	________	________

Answers: 1. mouth, then small intestine, amylase, smaller polysaccharides, maltose, glucose
2. small intestine, maltase, glucose
3. fiber is not digested by humans, excreted
4. small intestine, lipases, fatty acids and glycerol
5. stomach, then small intestine, pepsin, peptidases, polypeptides, and amino acids

VITAMINS (17.6)

I. True or False

1. ____ Vitamins are needed in small amounts by the body.
2. ____ Vitamins are inorganic compounds.
3. ____ Vitamins are secreted by various glands in the body.
4. ____ Vitamins are carried through the body by the bloodstream.
5. ____ Vitamins may act as coenzymes in metabolic reactions.
6. ____ Illness can occur if vitamins are not present in quantities required for good health.
7. ____ Illness can occur if fat-soluble vitamins are used in large quantities.

Answers: 1. T 2. F 3. F 4. T 5. T 6. T 7. T

J. Match the vitamins in the following with the statements below:

retinol (A)	thiamine (B_1)	niacin
cobalamine(B_{12})	ascorbic acid (C)	calciferol (D)
α-tocopherol (E)	phylloquinone (K)	

1. ____ The sunshine vitamin.
2. ____ A deficiency of this vitamin increases the blood clotting time.
3. ____ A vitamin that is required for night vision.
4. ____ A vitamin also known as ascorbic acid.
5. ____ A vitamin that contains cobalt.
6. ____ Low levels of this vitamin result in a condition called pellagra.
7. ____ Too little of this vitamin can lead to beriberi.
8. ____ This vitamin helps prevent oxidation of fatty acids in cell membranes.
9. ____ Without this vitamin, rickets may occur.
10. ____ Scurvy may indicate a deficiency of this vitamin.

Answers: 1. D 2. K 3. A 4. C 5. B_{12} 6. niacin 7. B_1 8. E 9. D 10. C

HORMONES (17.7)

K. Identify the hormone associated with each of the following statements:

A. insulin B. vasopressin C. oxytocin D. glucagon
E. estrogen F. parathyroid G. growth hormone H. testosterone

1._____ Stimulates the production of milk.

2._____ Increases the rate of carbohydrate metabolism.

3._____ Regulates calcium levels in the blood.

4._____ Causes development of secondary sex characteristics in females.

5._____ Stimulates reabsorption of water by the kidneys.

6._____ Promotes the growth of cells in the body.

7._____ Increases the blood level of glucose.

8._____ Stimulates the secondary sex characteristics in males.

Answers: 1. C 2. A 3. F 4. E 5. B 6. G 7. D 8. H

L. What hormone imbalance might be responsible for the following conditions?

A. lack of insulin B. too much growth hormone
C. too little growth hormone D. low testosterone
E. low estrogen F. too much insulin
G. low aldosterone H. high parathyroid

1. ____ A person who is 8-feet tall.

2. _____High sodium levels in urine.

3. ____ Failure of a 30-year-old male to develop secondary sex characteristics.

4. ____ Failure of a 20 year female to develop secondary sex characteristics.

5. ____ High blood glucose.

6. ____ Gigantism.

7. ____ Dwarfism

8. ____ Osteoporosis from loss of bone calcium.

Answers: 1. B 2. G 3. D 4. E 5. A 6. B 7. C 8. H

SELF-EVALUATION TEST
ENZYMES, VITAMINS, AND HORMONES

1. Enzymes are
   A. biological catalysts.
   B. polysaccharides.
   C. insoluble in water.
   D. always present in the mouth, stomach and small intestine.
   E. named with an "ose" ending.

Classify the enzymes described in questions 2, 3, 4 and 5 as
A. simple B. conjugated

2.____ An enzyme that yields amino acids and a glucose molecule on analysis.

3.____ An enzyme consisting of protein only.

4.____ An enzyme requiring zinc ion for activation.

5.____ An enzyme containing vitamin K.

For problems 6,7,8,9, and 10, select answers from the following:
(E = enzyme; S = substrate; P = product)

A. S $\longrightarrow$ P B. EP $\longrightarrow$ E + P C. E + S $\longrightarrow$ ES D. ES $\longrightarrow$ EP E. EP $\longrightarrow$ ES

6. ____ The enzymatic reaction occurring at the active site.

7. ____ The release of product from the enzyme.

8. ____ The first step in the lock-and-key theory of enzyme action.

9. ____ The formation of the enzyme-substrate complex.

10.____ The final step in the lock-and-key theory of enzyme action.

In problems 11,12,13,14, and 15, match the names of enzymes with a reaction they each catalyze.
A. decarboxylase B. isomerase C. dehydrogenase D. lipase E. sucrase

11.____ $NH_2\overset{OH}{\overset{|}{C}}HCOOH \longrightarrow NH_2\overset{O}{\overset{\|}{C}}COOH$

12.____ sucrose + $H_2O \longrightarrow$ glucose and fructose

13.____ $CH_3\overset{O}{\overset{\|}{C}}COOH \longrightarrow CH_3COOH + CO_2$

14.____ fructose $\longrightarrow$ glucose

15.____ triglyceride + $3H_2O \longrightarrow$ fatty acids and glycerol

For problems 16-20, select your answers from

A. Increases the rate of reaction
B. Decreases the rate of reaction
C. Denatures the enzyme and no reaction occurs.

What is the effect of each of the following upon the rate of an enzyme-catalyzed reaction?

16. ____ Setting the reaction tube in a beaker of water at 100°C.

17. ____ Adding substrate to the reaction vessel.

18. ____ Running the reaction at 10°C.

19. ____ Adding ethanol to the reaction system.

20. ____ Adjusting the pH to optimum pH.

For problems 21-25, identify the description of inhibition as

A. competitive B. noncompetitive C. irreversible

21. ____ An alteration in the conformation of the enzyme.

22. ____ A molecule closely resembling the substrate interferes with activity.

23. ____ The inhibition can be reversed by increasing substrate concentration.

24. ____ Activity is lost with very high temperatures.

25. ____ The inhibition is not affected by increased substrate concentration.

For problems 26-30, identify each vitamin as

W. water soluble F. fat soluble

26. ____ vitamin E

27. ____ vitamin A

28. ____ vitamin C

29. ____ vitamin D

30. ____ vitamin K

For problems 31-35, select answers from

A. retinol, vitamin A   B. niacin   C. ascorbic acid, vitamin C
D. calciferol, vitamin D   E. phylloquinone, vitamin K

31. ____Functions in the tissues of the body as the coenzymes NAD and NADH.

32. ____Needed for the formation of rhodopsin, a component of the visual cycle.

33. ____Its deficiency leads to prolonged bleeding times.

34. ____A fat soluble vitamin needed for the absorption of calcium from the intestinal tract.

35. ____Its precursors are the carotenes found in yellow or red vegetables.

For problems 36-40, select your answers from the following:

A. starch   B. fat   C. protein   D. sucrose   E. lactose

36. ___Digestion begins with amylase in the saliva.

37. ___The first step in digestion requires emulsification with bile salts.

38. ___Digestion begins in the stomach at pH 1-2.

39. ___Digestion occurs in the small intestine aided by the enzyme lactase.

40. ___Final digestive processes occur in the small intestine through the action of peptidases such as trypsin.

True or False:

41. ____ Hormones are most likely to be amino acids, peptides, or steroids.

42. ____ Hormones are organic compounds.

43. ____ Hormones must be supplied by the diet.

44. ____ Hormones are used by the gland where they are produced.

45. ____ Some hormones activate a cyclic AMP system in the cells to cause an increase in cellular activity.

Select answers from the following:

A. insulin B. estrogen C. vasopressin D. parathyroid E. thyroxine

46.____ Stimulates the secondary sex characteristics in females.

47.____ A small protein that increases the rate of carbohydrate metabolism.

48.____ Stimulates reabsorption of water by the kidneys.

49.____ Increases metabolic rate, utilization of food, and protein synthesis.

50.____ Regulates the levels of calcium and phosphate in body fluids.

ANSWERS TO THE SELF-EVALUATION TEST

1. A	11. C	21. B	31. B	41. T
2. B	12. E	22. A	32. A	42. T
3. A	13. A	23. A	33. E	43. F
4. B	14. B	24. C	34. D	44. F
5. B	15. D	25. B	35. A	45. T
6. D	16. C	26. F	36. A	46. B
7. B	17. A	27. F	37. B	47. A
8. C	18. B	28. W	38. C	48. C
9. C	19. C	29. F	39. E	49. E
10. B	20. A	30. F	40. C	50. D

SCORING THE SELF-EVALUATION TEST

50 questions 2 points each 100 points

CHAPTER 18
NUCLEIC ACIDS AND HEREDITY

KEY CONCEPTS

1. The nucleic acids, DNA and RNA, are large molecules composed of many nucleotides. The nucleotides consist of a phosphate group, a ribose sugar, and four kinds of nitrogen bases. DNA contains the nitrogen bases, adenine, cytosine, guanine, and thymines. In RNA, uracil is found in place of thymine.

2. The DNA molecule consists of a double helix formed by sugar-phosphate strands on the outside with complementary pairs of nucleotides along the center of the helix. Adenine always pairs with thymine (A=T), and cytosine always pairs with guanine (C≡G).

3. During replication, copies of DNA are formed through the process of complementary base pairing. The strands of the double helix separate and each half picks up the appropriate complementary base to give two identical copies of the DNA.

4. Ribonucleic acid, or RNA, is a single-chain ribonucleic acid. It takes the forms of messenger RNA (mRNA), ribosomal RNA (rRNA), and transfer RNA (tRNA). In a process called transcription, mRNA is formed in the nucleus by complementary base pairing with a DNA template. In RNA, the complementary base of adenine is uracil (A=U).

5. Translation is the process of converting the information in the mRNA into the sequence of amino acids of a protein. Each triplet of bases along the mRNA, known as the genetic code, directs a particular amino acid into the peptide chain. Each codon can be identified by the tRNA molecules which pick up their specific amino acids.

6. In protein synthesis, a mRNA carrying the genetic message migrates to the ribosomes in the cytoplasm to direct the synthesis of protein. In the process of translation, each codon on the mRNA is attached to a tRNA-amino acid. A peptide bond forms between a growing peptide chain and the newly placed amino acid until the order of amino acids for the protein is complete.

7. To make efficient use of cellular materials, metabolic pathways are used only when the end products are needed in the cell. Some control models include enzyme induction in which a substrate induces the synthesis of its enzymes, and enzyme repression in which a sufficient quantity of end product turns off further synthesis.

8. Mutations occur when errors are made in the transmission of the base sequence of a DNA molecule to the genetic code in a mRNA. Such errors alter the amino acid order affecting the ability of the protein to function properly.

KEY WORDS

Using complete sentences, define the following key terms:

DNA

RNA

replication

transcription

translation

mutation

LEARNING EXERCISES

NUCLEIC ACIDS: DNA AND RNA (18.1)

A. 1. What are the chemical components found in the nucleotides of nucleic acids?

2. What are the names and abbreviations for the nitrogen bases in

DNA?______________________________

RNA?______________________________

3. What is the sugar in the nucleotides of

DNA?______________________________

RNA?______________________________

Answers: 1. sugar, phosphate, nitrogen base
2. DNA:adenine (A), thymine (T), guanine (G), cytosine (C)
RNA: adenine (A), uracil (U), guanine (G), cytosine (C)
3. DNA: deoxyribose RNA: ribose

B. Identify the nucleic acid (DNA or RNA) in which the following nucleotides would be found.

1. adenosine monophosphate ________
2. dCMP ________
3. deoxythymidine monophosphate ________
4. UMP ________
5. guanosine monophosphate ________

Answers: 1. RNA 2. DNA 3. DNA 4. RNA 5. RNA

COMPLEMENTARY BASE PAIRING IN DNA (18.2) AND DNA REPLICATION (18.3)

C. Complete the DNA section by writing the complementary strand:

1. A-T-G-C-T-T-G-G-C-T-C-C-
2. A-A-A-T-T-T-C-C-C-G-G-G
3. G-C-G-C-T-C-A-A-A-T-G-C

Answers: 1. T-A-C-G-A-A-C-C-G-A-G-G
2. T-T-T-A-A-A-G-G-G-C-C-C
3. C-G-C-G-A-G-T-T-T-A-C-G

RNA: STRUCTURE AND SYNTHESIS (18.4)

D. List the function of each RNA if the cell.

nucleic acid	*function*
mRNA	
tRNA	
rRNA	

Answers: mRNA: carries information from DNA to the ribosomes (cytoplasm) for the synthesis of a protein. tRNA: picks up specific amino acids for protein synthesis as directed by rRNA. rRNA: found in the ribosomes; function not certain, but involved in protein synthesis.

GENETIC CODE (18.5)

E. Indicate the amino acid coded for by the following mRNA codons.

1. U—U—U ____________
2. A—G—C ____________
3. G—G—A ____________
4. G—C—G ____________
5. C—C—A ____________
6. A—C—A ____________

Answers: 1. Phe 2. Ser 3. Gly 4. Ala 5. Pro 6. Thr

PROTEIN SYNTHESIS (18.6)

F. Write the mRNA that would form for the following section of DNA. For each codon in the mRNA, write the amino acid that would be placed in the protein by a tRNA.

1. DNA: —C—C—C—T—C—A—G—G—G—C—G—C—

mRNA: ______________________________

amino acid order: ______________________________

2. DNA: —A—T—A—G—C—C—T—T—T—G—G—C—A—A—C—

mRNA: ______________________________

amino acid order: ______________________________

Answers: 1. mRNA: G-G-G-A-G-U-C-C-C-G-C-G-; Gly-Ser-Pro-Ala
2. mRNA: U-A-U-C-G-G-A-A-A-C-C-G-U-U-G-; Tyr-Arg-Lys-Pro-Leu

CELLULAR CONTROL OF PROTEIN SYNTHESIS (18.7)

G. Match the following descriptions of cellular control with the terms.

A. repressor B. operon C. structural gene
D. enzyme repression E. enzyme induction

____1. The production of an enzyme caused by the appearance of a substrate.

____2. A unit formed by a structural gene and an operator gene.

____3. High levels of end product stops the production of the enzymes in that pathway.

____4. A protein that attaches to the operator gene and blocks the synthesis of protein.

____5. The portion of DNA that produces the mRNA for protein synthesis.

Answers: 1. E 2. B 3. D 4. A 5. C

MUTATIONS (18.8)

H. Consider the DNA template of —A—A—T—C—C—C—G—G—G—

1. Write the mRNA produced.

2. Write the amino acid order for the mRNA codons.

3. Suppose a point mutation replaces the thymine in the DNA template with a guanine. Write the mRNA it produces.

4. What is the new amino acid order?

5. Why is this effect referred to as a point mutation?

6. How is a point mutation differ from an insertion or deletion mutation?

Answers: 1. -U-U-A-G-G-G-C-C-C- 2. Leu-Gly-Pro
3. -U-U-C-G-G-G-C-C-C- 4. Phe-Gly-Pro
5. In a point mutation, only one codon is affected, and one amino acid substituted.
6. In an insertion or deletion mutation, all of the triplet codes after the mutation is affected, producing many changes in the amino acid order.

SELF-EVALUATION TEST
NUCLEIC ACIDS AND HEREDITY

1. A nucleotide contains
   A. a nitrogen base
   B. a nitrogen base and a sugar
   C. a phosphate and a sugar
   D. a nitrogen base and a deoxyribose
   E. a nitrogen base, a sugar, and a phosphate

2. The double helix in DNA is held together by
   A. hydrogen bonding
   B. ester linkages
   C. peptide bonds
   D. salt bridges
   E. disulfide bonds

3. The process of producing DNA in the nucleus is called
   A. complementation
   B. replication
   C. translation
   D. transcription
   E. mutation

4. Which occurs in RNA but ***NOT*** in DNA?
   A. thymine
   B. cytosine
   C. adenine
   D. phosphate
   E. uracil

5. Which molecule determines protein structure in protein synthesis?
   A. DNA
   B. mRNA
   C. tRNA
   D. rRNA
   E. ribosomes

6. Which type of molecule carries amino acids to the ribosomes?
   A. DNA
   B. mRNA
   C. tRNA
   D. rRNA
   E. protein

For questions 7-15, select answers from the following nucleic acids:

A. DNA B. mRNA C. tRNA D. rRNA

7._____Along with protein, it is a major component of the ribosomes.

8._____A double helix consisting of two chains of nucleotides held together by hydrogen bonds between nitrogen bases.

9._____A nucleic acid that uses deoxyribose as the sugar.

10._____A nucleic acid produced in the nucleus which migrates to the ribosomes to direct the formation of a protein.

11._____It can place the proper amino acid into the peptide chain.

12._____It has nitrogen bases of adenine, cytosine, guanine, and thymine.

13._____It contains the codons for the amino acid order.

14._____It contains a triplet called an anticodon loop.

15._____This nucleic acid is replicated during cellular division.

For questions 16-20, select answers from the following:

A. —A—G—C—C—T—A—
| | | | | |
—T—C—G—G—A—T—

B. —A—U—U—G—C—U—C—

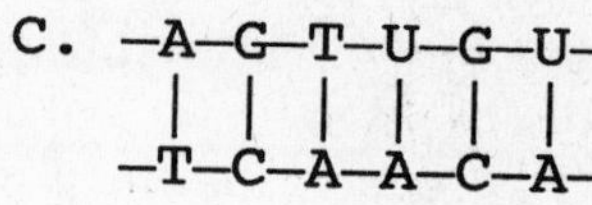

D. —G—U—A—

E. —A—T—G—T—A—T—

16._____A section of a mRNA.

17._____An impossible section of DNA.

18._____A codon.

19._____A section from a DNA molecule.

20._____A single strand that would not be possible for mRNA.

Read the following statements:

A. tRNA assembles the amino acids at the ribosomes.
B. DNA forms a complementary copy of itself called mRNA.
C. Protein is formed and breaks away.
D. tRNA picks up specific amino acids.
E. mRNA goes to the ribosomes.

Of the statements above, select the order in which they occur during protein synthesis.

21.____First step

22.____Second step

23.____Third step

24.____Fourth step

25.____Fifth step

For questions, 26-30, select your answers from the following:

A. mutation
B. enzyme induction
C. enzyme repression
D. operon
E. repressor

26.____A unit attaches to the operator gene and blocks the synthesis of a protein.

27.____An error in the transmission of the base sequence of DNA.

28.____A portion of a gene composed of the operating gene and the structural genes.

29.____A type of regulation whereby a substrate causes the synthesis of the enzymes necessary for its metabolism.

30.____The level of end product regulates the synthesis of the enzymes in that metabolic pathway.

ANSWERS TO THE SELF-EVALUATION TEST

1. E	6. C	11. C	16. B	21. B	26. E
2. A	7. D	12. A	17. C	22. E	27. A
3. B	8. A	13. B	18. D	23. D	28. D
4. E	9. A	14. C	19. A	24. A	29. B
5. A	10. B	15. A	20. E	25. C	30. C

SCORING THE SELF-EVALUATION TEST

questions 1-20	3 points each	60 points
questions 21-30	4 points each	40 points
	total	100 points

CHAPTER 19
METABOLIC PATHWAYS AND ENERGY PRODUCTION

KEY CONCEPTS

1. Energy in our cells is obtained from oxidation reactions of glucose, fatty acids, glycerol, and amino acids. When compounds such as glucose are oxidized, energy is released that is used to build molecules of ATP, the major energy source for many of the metabolic activities occurring in the cells.

2. ATP is produced in the mitochondria when NADH or $FADH_2$ bring hydrogen to the electron transport chain; 3 ATP molecules are produced for each molecule of NADH, and 2 molecules of ATP are produced for each molecule of $FADH_2$.

3. Glucose begins its oxidation by glycolysis which converts glucose to pyruvic acid.

4. The aerobic oxidation of pyruvic acid produced acetyl CoA, CO_2, water and 3 ATP molecules for each molecule of acetyl CoA. Since a molecule of glucose produces two molecules of pyruvic acid, a total of 6 ATP molecules result. Without oxygen, pyruvic acid is converted to lactic acid, a process that produces only 2 molecules of ATP for each molecule of glucose that reacts. In yeast, pyruvic acid undergoes fermentation (anaerobic) to produce energy.

5. Aerobic oxidation continues when acetyl CoA enters the citric acid cycle where an additional 12 ATP molecules are produced for each acetyl CoA, or a total of 24 molecules of ATP from the initial molecule of glucose.

6. The entire aerobic conversion of 1 molecules glucose to CO_2 and H_2O produces 36 molecules of ATP: 6 ATP molecules from glycolysis, 6 ATP molecules from the formation of acetyl CoA, and 24 ATP molecules from the oxidation reactions of 2 acetyl CoA in the citric acid cycle.

7. When blood glucose is not available as an energy source, fatty acids are oxidized to acetyl CoA units that enter the citric acid cycle to produce ATP energy.

8. Amino acids may be used as an energy source after the amino group is removed and an α-keto acid produced that is a component of the citric acid cycle.

KEY WORDS

Using complete sentences, write a definition for the following terms:

ATP

electron transport chain

glycolysis

citric acid cycle

beta (ß)-oxidation

LEARNING EXERCISES

ATP: THE ENERGY STOREHOUSE (19.1)

A. Fill in the blanks with following terms: ATP phosphate catabolic ADP adenine anabolic

In the cell, __________(1) processes extract energy by breaking large molecules down into smaller molecules. When energy is required to build new and larger molecules for the cells, the process is known as __________(2) reactions. The energy extracted is used to build a high-energy compound known as ______(3). This energy-rich molecule is composed of the purine base _________(4), a sugar, and three ____________(5) groups. Its hydrolysis yields _______(6) + P_i and 7.3 kcal of energy.

Answers: 1. catabolic 2. anabolic 3. ATP 4. adenine 5. phosphate 6. ADP

B. Write an equation that illustrates the hydrolysis of ATP.

Answer: ATP $\longrightarrow$ ADP + P_i + energy

THE ELECTRON TRANSPORT CHAIN (19.2)

C. Consider the components of the electron transport chain.

1. List the oxidized and reduced forms of the coenzymes NAD^+ and FAD.

   oxidized　　　　　　reduced

   ________ ________　　________ ________

2. List the oxidized and reduced forms of the electron carriers of flavin mononucleotide and coenzyme Q.

   oxidized　　　　　　reduced

   ________ ________　　________ ________

3. Write an equation for the transfer of hydrogen from $FMNH_2$ to Q.

4. What is the function of coenzyme Q in the electron transport chain?

5. Write the oxidized and reduced forms of cytochrome b.

   oxidized ____________　　reduced ____________

6. What are the end products of the electron transport chain?

7. How much ATP is produced when NAD^+ is the initial hydrogen acceptor?

8. How much ATP is produced when FAD is the initial hydrogen acceptor?

Answers:

1. oxidized: NAD^+ and FAD ; reduced: $NADH + H^+$ and $FADH_2$
2. oxidized: FMN and Q; reduced: $FMNH_2$ and QH_2
3. $FMNH_2 + Q \longrightarrow FMN + QH_2$
4. Coenzyme Q accepts hydrogen atoms from $FMNH_2$ or $FADH_2$. From QH_2, the hydrogen atoms are separated into protons and electrons with the electrons being passed on to the cytochromes, the electron acceptors in the chain.
5. $2\ \text{Cyt b}\ (Fe^{3+}) + 2\ e^- \longrightarrow 2\ \text{Cyt b}\ (Fe^{2+})$
6. $CO_2 + H_2O + ATP$ (energy)
7. 3 ATP for each NADH
8. 2 ATP for each $FADH_2$

GLYCOLYSIS: OXIDATION OF GLUCOSE (19.3)

D. Fill in the blanks with the following terms: aerobic, anaerobic, glucose, pyruvic acid

Glycolysis is an important series of reactions in the cell whereby a molecule of __________(1) is converted to two molecules of ____________(2). Under __________(3) conditions, a total of 6 ATP molecules are produced.

Answers: 1. glucose 2. pyruvic acid 3. aerobic

PATHWAYS FOR PYRUVIC ACID (19.4)

E. Fill in the blanks with the following terms

lactic acid	citric acid	NAD^+	fermentation
NADH	aerobic	anaerobic	acetyl CoA

When oxygen is available during glycolysis, the three-carbon pyruvic acid may be oxidized further by forming _____________(1) + CO_2. The hydrogen is removed by _________(2) to give __________(3) which in turn provides 3 ATP molecules via the electron transport chain. The acetyl CoA provides 12 more ATP molecules when it enters the _______________(4) cycle. Under ___________(5) conditions, pyruvic acid is reduced in the cytoplasm to ________(6) by NADH. In yeast, pyruvic acid is converted to ethanol by NADH, a process known as _______________(7).

Answers: 1. acetyl CoA 2. NAD^+ 3. NADH 4. citric acid 5. anaerobic 6. lactic acid 7. fermentation

CITRIC ACID CYCLE (19.5)

F. In each of the following steps of the citric acid cycle, indicate whether or not an oxidation is associated with the reaction. If so, give the coenzyme (NAD^+ or FAD) and the ATP produced.

Step in citric acid cycle	*oxidation*	*coenzyme*	*ATP*
1. acetyl CoA + oxaloacetic acid ⟶ citric acid	____	____	____
2. citric acid ⟶ isocitric acid	____	____	____
3. isocitric acid ⟶ α-ketoglutaric acid	____	____	____
4. α-ketoglutaric acid ⟶ succinyl CoA	____	____	____
5. succinyl CoA ⟶ succinic acid	____	____	____
6. succinic acid ⟶ fumaric acid	____	____	____
7. fumaric acid ⟶ malic acid	____	____	____
8. malic acid ⟶ oxaloacetic acid	____	____	____

Answers: 1. none 2. none 3. yes, NAD^+, 3 ATP 4. yes NAD^+, 3 ATP 5. no, but it does have a direct phosphorylation(GDP), 1 ATP 6. yes, FAD, 2 ATP 7. none 8. yes, NAD^+, 3 ATP

ENERGY PRODUCTION FROM GLUCOSE (19.6)

G. Complete the following:

substrate	*reaction*	*products*	*ATP*
1. acetyl CoA	citric acid cycle	________	____
2. pyruvic acid	oxidation	________	____
3. glucose	glycolysis(aerobic)	________	____
4. glucose	glycolysis(anaerobic)	________	____
5. glucose	complete oxidation	________	____

Answers: 1. $2CO_2$; 12 ATP 2. acetyl SCoA + CO_2; 3 ATP 3. 2 pyruvic acid; 6 ATP 4. 2 lactic acid; 2 ATP 5. $6CO_2$ + $6H_2O$; 36 ATP

METABOLIC PATHWAYS FOR FATTY ACIDS (19.7)

H. Decanoic acid is a 10-carbon fatty acid.

1. How much ATP is needed for activation?

2. How many NADH AND $FADH_2$ are produced during ß-oxidation?

3. How many acetyl CoA units will result?

4. What is the total ATP produced from the electron transport chain and the citric acid cycle?

Answers: 1. 2 ATP 2. 4 turns of cycle produces 4 NADH and 4 $FADH_2$. 3. 5 acetyl CoA result 4. 4 NADH x 3 ATP = 12 ATP; 4 $FADH_2$ x 2 ATP = 8 ATP; 5 acetyl CoA x 12 ATP = 60 ATP; total ATP = 12 ATP + 8 ATP + 60 ATP - 2 (activation) = 78 ATP

METABOLIC PATHWAYS FOR AMINO ACIDS (19. 8)

I. Explain the following steps in metabolic pathways of amino acids.

1. What is the first step in preparing an amino acid for energy production?

2. What happens to the α-keto acids produced?

3. How can the intermediates of the citric acid cycle be used to prepare some of the nonessential amino acids?

Answers: 1. The amino group is removed through transamination. 2. The resulting α-keto acids are used in or converted to α-keto acids that enter the citric acid cycle where they are oxidized during energy production. 3. The α-keto acids of the citric acid cycle may be transaminated by transferring an amino group from glutamic acid to provide the nonessential amino acids.

SELF-EVALUATION TEST

METABOLIC PATHWAYS AND ENERGY PRODUCTION

1. The main function of the mitochondria is
   A. energy release
   B. protein synthesis
   C. glycolysis
   D. genetic instructions
   E. waste disposal

2. ATP serves as a
   A. nucleotide unit in RNA and DNA
   B. end product of glycogenolysis
   C. end product of transamination
   D. enzyme
   E. energy storage molecule

3. The end products of the electron transport chain are
   A. H_2O + ATP
   B. $CO_2 + H_2O$
   C. $NH_3 + CO_2 + H_2O$
   D. $H_2 + O_2$
   E. urea (NH_2CONH_2)

4. Glycolysis
   A. requires oxygen for the catabolism of glucose.
   B. represents the aerobic sequence for glucose anabolism and ATP production.
   C. represents the splitting off of glucose residues from glycogen.
   D. represent the anaerobic catabolism of glucose to lactic acid and ATP production
   E. produces acetyl units and ATP as end products

5. Which does ***NOT*** appear in the glycolysis pathway?
   A. dihydroxyacetone phosphate
   B. pyruvic acid
   C. NAD^+
   D. acetyl CoA
   E. lactic acid

6. Which is true of the citric acid cycle?
   A. Most of the potential energy stored in foods is released.
   B. Succinic acid combines with acetyl units to form citric acid.
   C. NAD^+ is the only hydrogen acceptor used.
   D. The electron transport chain is not connected to the citric acid cycle.
   E. Only carbohydrates are metabolized by this cycle.

7. Which is found in the citric acid cycle?
   A. glucose
   B. α-ketoglutaric acid
   C. lactic acid
   D. glutamic acid
   E. stearic acid

8. The electron transport chain
   A. produces most of the body's ATP.
   B. carries oxygen to the cells.
   C. produces $CO_2 + H_2O$.
   D. is only involved in the citric acid cycle.
   E. operates during fermentation.

9. Which is ***NOT*** found in the ß-oxidation pathway?
   A. fatty acids B. NAD^+ C. acetyl CoA D. glycerol E. FAD

10. The process of transamination
    A. is part of the citric acid cycle.
    B. converts α-amino acids to ß-keto acids.
    C. produces new amino acids.
    D. is not used in the metabolism of amino acids.
    E. is part of ß-oxidation of fats.

For questions 11-20, match the statements with the correct metabolic process:

11.____ Reactions in the presence of oxygen.

12.____ Storage form of energy in the cell.

13.____ A group of electron acceptors in the respiratory chain.

14.____ A coenzyme.

15.____ Preparation of fatty acids for the citric acid cycle.

16.____ Reaction conditions without oxygen.

17.____ A series of reactions whereby glucose is converted to pyruvic acid.

18.____ The conversion of glucose to glycogen.

19.____ A series of reactions that converts acetyl CoA to CO_2 and water.

20.____ The preparation of amino acids for use as an energy source.

A. ATP
B. glycolysis
C. aerobic
D. anaerobic
E. cytochromes
F. glycogenesis
G. oxidative deamination
H. citric acid cycle
I. ß-oxidation
J. NAD^+